Machining of Metal Matrix Composites

J. Paulo Davim
Editor

Machining of Metal Matrix Composites

Springer

Prof. Dr. J. Paulo Davim
Department of Mechanical Engineering
University of Aveiro
Campus Santiago
3810-193 Aveiro
Portugal
e-mail: pdavim@ua.pt

ISBN 978-0-85729-937-6 e-ISBN 978-0-85729-938-3
DOI 10.1007/978-0-85729-938-3
Springer London Dordrecht Heidelberg New York

British Library Cataloguing in Publication Data
A catalogue record for this book is available from the British Library

Library of Congress Control Number: 2011936534

Cover design: eStudio Calamar S.L.

Printed on acid-free paper

Springer is part of Springer Science+Business Media (www.springer.com)

Preface

In recent years, the utilization of metal-matrix composites (MMCs) has increased in various areas of science and technology due to their special mechanical and physical properties. MMCs, particularly aluminium-based composites have a high strength-to-weight ratio, high stiffness, lower thermal expansion coefficient, high thermal conductivity as well as corrosion and wear resistance. Therefore, MMCs have the potential to replace conventional materials in various fields of application such as automotive, aeronautical and aerospace as well as in others advanced industries. As result of these potential applications, there exist a great necessity to understand the problems associates with the machining of these composites. Machining MMCs is a rather complex task owing to it is heterogeneity and to the fact that reinforcements are extremely abrasive and responsible for complex deformation behavior, high tool wear and inferior surface finish.

Chapter 1 of this book provides the mechanics and modelling of chip formation in machining of MMC. Chapter 2 is dedicated to surface integrity when machining metal-matrix composites. Chapter 3 described machinability aspects of metal-matrix composites. Chapter 4 contains information on traditional machining processes and Chap. 5 is dedicated to grinding of metal-matrix composites. Chapter 6 described dry cutting of SiC particulates reinforced metal-matrix composite. Finally, Chap. 7 is dedicated to computational methods and optimization in machining of metal-matrix composites.

The present book can be used as a text book for final undergraduate engineering course or as a topic on manufacturing at the postgraduate level. Also, this book can serve as a useful reference for academicians, manufacturing and materials researchers, manufacturers, materials and mechanical engineers, professionals in composites and related industries. The interest of scientific in this book is evident for many important centers of the research, laboratories and universities throughout the world. Therefore, it is hoped that this book will inspire and enthuse other researches for this field of the machining science and technology.

The Editor acknowledges Springer for this opportunity and for their enthusiastic and professional support. Finally, I would like to thank all the chapter authors for their availability for this work.

March 2011 J. Paulo Davim

Contents

Contributors

Prof. Yung C. Shin School of Mechanical Engineering, Purdue University, West Lafayette, IN, USA, e-mail: shin@purdue.edu

Dr. Chinmaya Dandekar School of Mechanical Engineering, Purdue University, West Lafayette, IN, USA

Prof. Abdul B. Sadat Industrial and Manufacturing Department, California State Polytechnic University, Pomona, CA 91768, USA, e-mail: absadat@csupomona.edu

Prof. Antoniomaria Di Ilio Università dell'Aquila, Piazza V. Rivera 1, 67100 L'Aquila, Italy, e-mail: antoniomaria.diilio@univaq.it

Prof. Alfonso Paoletti Università dell'Aquila, Piazza V. Rivera 1, 67100 L'Aquila, Italy, e-mail: alfonso.paoletti@univaq.it

Prof. H. A. Kishawy Faculty of Engineering and Applied Science, University of Ontario Institute of Technology, 2000 Simcoe Street North, Oshawa, ON L1H 7K4, Canada, e-mail: Hossam.Kishawy@uoit.ca

Dr. S. Kannan Manufacturing Technology Group, Rolls-Royce plc, Wilmore Road, Derby DE24 8BJ, UK, e-mail: jeshtasathish@yahoo.co.in

Dr. G. Parker Faculty of Engineering and Applied Science, University of Ontario Institute of Technology, 2000 Simcoe Street North, Oshawa, ON L1H 7K4, Canada, e-mail: Grant.Parker@uoit.ca

Dr. B. Anand Ronald Department of Mechanical Engineering, SSN College of Engineering, Kalavakkam 603110, India, e-mail: anandronald@rediffmail.com

Prof. L. Vijayaraghavan Department of Mechanical Engineering, Indian Institute of Technology Madras, Chennai 600036, India, e-mail: lvijay@iitm.ac.in

Prof. R. Krishnamurthy Department of Mechanical Engineering, Indian Institute of Technology Madras, Chennai 600036, India, e-mail: ramalingamkrishnamurthy@gmail.com

Dr. Arnaud Kremer LMPF-EA 4106, Arts et Metiers ParisTech, rue Saint Dominique, 51000 Chalons-en-Champagne, France, e-mail: arnaud.kremer@ensam.eu

Prof. Mohamed El Mansori LMPF-EA 4106, Arts et Metiers ParisTech, rue Saint Dominique, 51000 Chalons-en-Champagne, France, e-mail: mohamed.elmansori@ensam.eu

Prof. V. N. Gaitonde Department of Industrial and Production Engineering, B. V. B. College of Engineering and Technology, Hubli, Karnataka 580031, India, e-mail: gaitondevn@yahoo.co.in

Prof. S. R. Karnik Department of Electrical and Electronic Engineering, B. V. B. College of Engineering and Technology, Hubli, Karnataka 580031, India, e-mail: karniksr@yahoo.com

Prof. J. Paulo Davim Department of Mechanical Engineering, University of Aveiro, Campus Santiago, 3810-193 Aveiro, Portugal, e-mail: pdavim@ua.pt

Chapter 1
Mechanics and Modeling of Chip Formation in Machining of MMC

Yung C. Shin and Chinmaya Dandekar

Abstract Metal matrix composites (MMCs) offer high strength-to-weight ratio, high stiffness and good damage resistance over a wide range of operating conditions, making them an attractive option in replacing conventional materials for many engineering applications. Typically the metal matrix materials of MMCs are aluminum alloys, titanium alloys, copper alloys and magnesium alloys, while the reinforcement materials are silicon carbide, aluminum oxide, boron carbide, graphite etc. in the form of fibers, whiskers and particles. This chapter covers the mechanics of chip formation during machining of MMCs and various modeling techniques. Especially, modeling techniques dealing with cutting force, chip morphology, temperature and subsurface damage are covered.

1.1 Introduction

Metal matrix composites (MMCs) offer high strength to weight ratio, high stiffness and good damage resistance over a wide range of operating conditions, making them an attractive option in replacing conventional materials for many engineering applications. Typically the metal matrix materials of MMCs are aluminum alloys, titanium alloys, copper alloys and magnesium alloys, while the reinforcement materials are silicon carbide, aluminum oxide, boron carbide, graphite etc. in the form of fibers, whiskers and particles. Probably the single most important difference between fiber reinforced and particulate composites or conventional metallic materials is the directionality of properties. Particulate composites and conventional metallic materials are isotropic, while the fiber reinforced composites are generally anisotropic. Particulate reinforced composites offer higher ductility and

Y. C. Shin (✉) · C. Dandekar
School of Mechanical Engineering, Purdue University, West Lafayette, IN, USA
e-mail: shin@purdue.edu

J. P. Davim (ed.), *Machining of Metal Matrix Composites*,
DOI: 10.1007/978-0-85729-938-3_1,

their isotropic nature as compared to fiber reinforced composites makes them an attractive alternative.

Machining of particulate reinforced MMC has been extensively studied experimentally in the past, while studies on the machining of fiber-reinforced MMCs is limited. MMCs are shown to cause excessive tool wear, which in turn induces such damage phenomena as fiber pullout, particle fracture, delamination and debonding at the fiber or particle and matrix interface. The parameters that are the major contributors to the machinability of these composites are the reinforcement type and orientation, tool type and geometry and the machining parameters. Although MMCs are generally processed near-net shape, subsequent machining operations are inevitable.

The methods used in studying the machining of composites have been diverse, and the investigations can be generally divided into three categories: experimental studies focusing on the macro/microscopic machinability of composites, simple modeling using conventional cutting mechanics, and numerical simulations that treat a composite as a macroscopically anisotropic material or concentrate on the reinforcement–matrix interaction microscopically. The macroscopic models normally ignore many fundamental characteristics of composites subjected to cutting and usually cannot be well integrated with the cutting mechanics, while those focusing on the micro-effects, including the analysis using the finite element method, are tedious to implement. A sensible way seems to combine the merits of these methods to develop realistic models that not only depict the material removal mechanisms in cutting, but also provide simple, analytical solutions for applications.

This chapter deals with understanding the mechanics of chip formation in machining of MMC. Optimization of machining parameters to achieve a better surface finish, reduced damage and maximum tool life is highly desirable. Understanding the mechanics of cutting would assist in selecting the optimum machining parameters so as to improve the machinability of these composites.

1.2 Machining of Particulate Reinforced Metal Matrix Composites

Most of the research related to machining of particulate reinforced MMC is attributed to turning and has been extensively studied experimentally in the past to assess the cutting forces, cutting temperature, the attendant tool wear, surface roughness and sub-surface damage. From the available literature on machining of MMCs it is obvious that the reinforcement material, type of reinforcement (particle or whisker), volume fraction of the reinforcement, and matrix properties as well as the distribution of these particles in the matrix are the factors that affect the overall machinability of these composites. The most commonly used tool material is polycrystalline diamond (PCD) [1–7], although cubic boron nitride (CBN), alumina, silicon nitride and tungsten carbide (WC) tooling are also used as cutting materials. Cutting speed, feed and depth of cut in machining of particulate MMCs have a similar effect on tool life and surface finish to that of machining metals although some differences are noticeable due to the ceramic particles. The ceramic-reinforced particles tend to dislodge from the matrix and roll in front of the cutting tool, thereby plowing through the machined surface and generating grooves on it [8, 9].

1.2.1 Effect of Cutting Speed

In most cases, cutting speed does not significantly influence the cutting forces [10]. There are some contradictory reports on the effect of cutting speed on the cutting forces. During machining of MMCs a built-up edge (BUE) has been observed by many researchers while machining these composites at low cutting speeds [9, 11, 12]. Due to the BUE the cutting force at low cutting speeds is lower than the cutting force observed at higher cutting speeds. This phenomenon can be attributed to either higher tool wear at high cutting speeds or the presence of BUE. The presence of a BUE increases the actual rake angle of the tool resulting in a lower cutting force. There are some studies which have shown a decrease in the cutting forces with an increase in the cutting speed [13, 14]. In the study conducted by Manna and Bhattacharya [14], the influence of the cutting speed on the feed force and cutting force during turning of an Al/SiC composite was measured. The experimental results showed that the feed force and the cutting force decreased with an increase in the cutting speed.

The tool life decreases while the surface finish improves only slightly with an increase in cutting speed, since the tool temperature increases with cutting speed, thereby softening the tool material and consequently accelerating the diffusion wear [3, 15, 16]. Overall, the variation of surface roughness with cutting speed is not significant as the surface roughness is dominated by the size of reinforcement and the feed [3, 16, 17]. In terms of tool life, Manna and Bhattacharya [9, 11] conducted studies using carbide tooling for machining of an Al/SiC composite and observed that the flank wear increased 2.5–3 times for an increase in cutting speed from 60 to 180 m/min. Another observation of flank wear variation with cutting speed is the very rapid increase in flank wear at cutting speeds above 100 m/min and hence cutting speed range of 60–100 m/min was suggested for machining of these composites. Ozben et al. [12] and Joshi et al. [18] both machined an aluminum matrix reinforced with SiC particles and observed that the cutting speed was one of the dominant factors in limiting the machinability of the composite.

1.2.2 Effect of Feed

Feed has a significant effect on the cutting forces in that the cutting forces rise considerably with an increase in the feed [10, 19]. There are many force prediction models available in the literature for machining of MMCs. For instance, Kishawy et al. [20] developed an energy-based analytical model to predict the forces in orthogonal cutting of an MMC using a ceramic tool at a low cutting speed, while Pramanik et al. [19] developed a mechanics-based model for prediction of the cutting forces based on the mechanisms of chip formation, and the presence of the matrix and particle. Feed, on the other hand, negatively influences the surface roughness, where the surface finish deteriorates with an increase in feed [3, 21]. Furthermore feed has the largest effect on the damage observed in the sub-surface [8, 22, 23], where larger feed results in more damage and also greater damage depth into the material. El-Gallab and Sklad [8, 23] concluded that the failure in the composite initiates along the voids generated around the SiC particles due to the high cutting forces observed at higher feeds. The voids join up to form micro-cracks and subsequent fracture along the shear band. On the other hand feed tends to have less influence on the tool wear. A high feed can reduce the tool-wear rate due to the improvement in the conduction of heat from the cutting zone to the workpiece [15]. Feed increases the flank wear but only marginally as compared to cutting speed. At a cutting speed of 60 m/min increasing the feed three folds increased the flank wear 1.6 times, while increasing the speed three folds at a feed of 0.35 mm/rev the flank wear increased three times [9, 24].

1.2.3 Effect of Depth of Cut

Depth of cut has a negative effect on the surface finish and the sub-surface damage. An increase in depth of cut decreases the quality of the surface finish and the sub-surface damage. Chambers [25] conducted a study on machining of a 15% by volume fraction of SiC in A356 aluminum alloy and concluded that the depth of cut did not significantly alter the tool life, with tool life decreasing with an increase in the depth of cut. Although the effect of depth of cut on tool wear is not significant, it has a stronger effect on the tool wear as compared to the feed as shown in machining of an Al/SiCp/15% composite with uncoated tungsten carbide tools [9, 15]. Additionally, an increase in the depth of cut increases the machining forces during the machining of MMCs.

1.2.4 Effect of the Reinforcement

Absent in machining of homogenous materials, the presence of the reinforcement affects the machinability of composites substantially. The hard ceramic particles in the matrix cause numerous problems, especially the excessive tool wear. The size and the percentage volume fraction of the reinforcement play a significant role on the machinability of composites. An expected result is the progression of tool wear and surface finish is highly dependent on the average size and volume fraction of the particles. Ciftci et al. [26] machined an Al/SiCp composite with SiC particle size of 30, 45 and 110 μm and a reinforcement of 16% volume using both coated and uncoated carbide tools. The authors reported that the tool wear and the surface finish are negatively affected by the particle size. This observation was further substantiated by Kannan et al. [15] while machining a composite with 10% by volume fraction of alumina particles in an Al 6061 matrix, with average particle size of 9.5, 17, 20 and 25 μm. An increase in particle volume fraction also results in increased tool wear and subsequently affects the surface finish of the machined workpiece. Higher tool wear is the result of the hard ceramic particles seen at a higher frequency by the cutting tool [3, 15, 26].

Similarly, Ozben et al. [12] machined an aluminum matrix reinforced with SiC particles in 5, 10 and 15% by volume fraction and observed that the cutting speed and percentage volume fraction of the particles were the dominant factors in limiting the machinability of the composite. Joshi et al. [18] studied the effect of feed (0.084–0.17 mm/rev), cutting speed (22–88 m/min), tool inclination angle (15 and 45°) and percentage volume fraction of SiC particles in aluminum (10 and 30%) on machining of the MMC with a carbide tool and arrived at an empirical relationship between flank wear and cutting time as a function of the aforementioned parameters. The authors concluded that the cutting speed and the percentage volume fraction of the particles had the most significant effect on the tool life.

1.2.5 Tooling

PCD diamond tools are the most preferred, while carbide tools are preferred over ceramic tools [1–4, 6, 7, 27]. In case of carbide tooling low-cutting speeds and high-feed rates are utilized to maximize the tool life [15, 28]. High tool wear observed while machining of these composites is generally associated with carbide tooling. At higher cutting speeds (>350 m/min), the carbide tool demonstrates catastrophic failure and hence in the literature cutting speed is generally limited up to 300 m/min [9, 11, 12, 18, 25, 29–31]. In other tool materials, Tomac and Tonnessen [28] compared the performance of chemical vapor deposition (CVD) coatings of TiN, TiCN and Al_2O_3 and concluded that the inserts with TiN coating performed the best in maximizing the tool life. To improve the tool life in carbide tools, Manna and Bhattacharya [9] machined at cutting conditions that sustained a stable built-up edge (BUE) so as to protect the cutting tool. To minimize the surface roughness and sub-surface damage PCD tools are preferred since the wear rate associated with them is the lowest among available tool materials. Although PCD tools are used for machining Al/SiC composites, the high cost associated with them limits their use [3–5, 16, 27].

1.3 Machining of Fiber Reinforced Metal Matrix Composites

Ceramic fiber reinforced MMC have seldom been machined with conventional machining methods. The fibers can be either short or long and continuous as governed by their application. The reinforcements enhance the properties of the metal matrix by increasing the fracture toughness, resistance to high temperatures, strength and damage tolerance. The composite properties are highly dependent on the type of reinforcement as the mode of failure will differ. Continuous fiber reinforcements are stiffer than particulate or whisker reinforcements [32] in the fiber direction. Similar to machining of monolithic ceramics, continuous fiber MMCs are generally not machined using conventional machining techniques like milling and turning due to the hardness of the constituent fibers. Fibers present in the metal matrix pose another problem for machining of MMCs as any fiber breakage or pullout causes a reduction in the material properties. Furthermore silicon carbide (SiC) fibers and boron nitride interface are susceptible to oxidation and hence care must be taken during machining. Komanduri [33] reported that in machining of a glass reinforced with continuous fibers of silicon carbide no cutting tool material could achieve respectable tool life. In conventional machining tests, Varadarajan et al. [34] studied the machinability characteristics of a low-volume fraction (25%), random aluminosilicate fiber reinforced aluminum composite with two tool materials; coated carbides and polycrystalline cubic boron nitride (PCBN). The results indicated that PCBN tools outperformed coated carbides in terms of tool wear and surface finish. In another study on machining of short-fiber reinforced metal matrix composite, Weinert and Lange [35] assessed the machinability of a 20% by volume fraction of Al_2O_3 fiber in a magnesium alloy matrix, wherein milling was carried out with a PCD tool, and the authors suggested the use of moderate cutting speeds and high feed to offset the high tool wear. Machining experiments were conducted by Dandekar and Shin [36] on an Al-2%Cu aluminum matrix composite reinforced with 62% by volume fraction alumina fibers (Al-2%Cu/Al_2O_3). The cutting speed was 30 m/min with feed of 0.02 mm/rev and 0.5 mm depth of cut. The cutting tool material is a PCD tool with a tool rake angle of 5°. Damage in the composite was measured and it was characterized in this case through excessive fiber breakage below the cutting plane. Due to the brittle nature of the alumina fiber there was observable damage below the cutting plane. Damage was observed in the form of debonding between the fiber and the matrix, microcracking of fibers and fiber pullout.

The excessive tool wear and damage associated with machining of long-fiber reinforced MMC results in the process being uneconomical. It is clear that the presence of reinforcement makes MMCs different from monolithic materials due to incorporation of its superior physical properties. In addition, the amount and type of reinforcement introduce different properties in the strength and toughness of composites. Higher fiber/particulate reinforcement results in a reduction in the ductility of MMCs, causing harsh machining conditions.

1.4 Mechanics of Machining MMC's

Machining of MMCs can be classified into two major categories: a particulate reinforced and (b) fiber reinforced. Depending on the type of reinforcement, the cutting mechanics differ considerably. It is therefore clear that the tool–

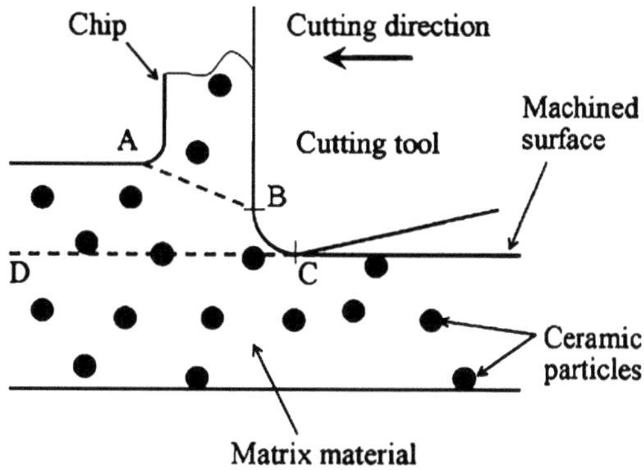

Fig. 1.1 Cutting of an MMC (*source*: Zhang [41], with permission from Elsevier)

reinforcement–matrix interactions play a significant role in the machinability of MMC's and affect the surface roughness, cutting forces, tool wear and the sub-surface damage.

1.4.1 Analytical Machining Model to Predict Cutting Forces

In terms of analytical modeling to predict the cutting forces, a number of authors have tackled this problem [19, 20, 37]. Kishawy et al. [20] were among the first to propose an analytical model for prediction of cutting forces. The model is an energy-based method, where the energy is estimated based on the deformation of the primary and secondary zones and the fracture and displacement of the reinforcement. Although the model was successful in predicting the cutting forces, it was based on two assumptions: the energy in the secondary deformation zone was one-third of that of the primary deformation zone and the initial and final crack lengths of the ceramic particles were 1 μm. The first assumption was based on the results obtained for machining of steel while no justification was made for the second assumption. Pramanik et al. [19] and Davim [37] both developed the cutting force model based on Merchant's orthogonal machining model. The major difference in their models is that Pramanik et al. [19] explicitly treat the effects of particles in their model as opposed to treating the composite as an equivalent homogenous material in the model proposed by Davim [37].

In the model developed by Pramanik et al. [19], the cutting force was predicted by splitting the total force into (a) the chip formation force, (b) the plowing force and (c) the particle fracture force. The chip formation force was obtained using Merchant theory [38], while the plowing and particle fracture forces were obtained with the aid of the slip line field theory of plasticity [39] and the Griffith theory of fracture respectively. Figure 1.1 shows a schematic of orthogonal machining of a particulate MMCs. According to Pramanik et al. [19] the mechanism of chip formation is due to shearing along the shear plane AB marked in Fig. 1.1. This assumption allows the chip formation mechanism to be similar to orthogonal machining of a monolithic material with a sharp tool. The plowing force (plastic zone with no chip) is due to the material deformation and the particle displacement, a result of the tool edge radius, as shown as BC in Fig. 1.1. Finally the force due to the particle fracture is accounted for along the line CD. This assumption is based on experimental observations of Yan and Zhang [40].

The total cutting force is therefore assumed to be a superposition of the cutting forces due to the individual contributions from chip formation, plowing and particle fracture as shown in Eqs. 1.1 and 1.2 for the total force in the cutting direction (F_C) and the thrust direction (F_T), respectively as:

$$F_C = F_{Cc} + F_{Cp} + F_{Cf} \tag{1.1}$$

$$F_T = F_{Tc} + F_{Tp} + F_{Tf} \tag{1.2}$$

where F_{Cc} and F_{Tc} are the forces due to chip formation, F_{Cp} and F_{Tp} are the forces due to plowing and F_{Cf} and F_{Tf} are the forces due to particle fracture. The force components for them are obtained as shown in Eqs. 1.3–1.7.

$$F_{Cc} = \tau_s A_c \frac{\cos(\beta - \gamma)}{\sin\phi \cos(\phi + \beta - \gamma)} \tag{1.3}$$

$$F_{Tc} = \tau_s A_c \frac{\sin(\beta - \gamma)}{\sin\phi \cos(\phi + \beta - \gamma)} \tag{1.4}$$

$$F_{Cp} = \tau_{sm} l r_n \tan\left(\frac{\pi}{4} + \frac{\gamma}{2}\right) \tag{1.5}$$

$$F_{Tp} = \tau_{sm} l r_n \left(1 + \frac{\pi}{2}\right) \tan\left(\frac{\pi}{4} + \frac{\gamma}{2}\right) \tag{1.6}$$

$$F_{Cf} = F_{Tf} = \left(\frac{\mu_g l}{L}\right) \tan\delta \tag{1.7}$$

where A_c is the cross-sectional area of the cut, τ_s is the shear strength of the MMC, τ_{sm} is the shear strength of the matrix, r_n is the edge radius, β is the angle of friction, γ is the tool rake angle, ϕ is the shear angle, δ is the resultant cutting force angel, L is the cutting distance, μ_g is the average fracture energy per unit cutting edge length and l is the active cutting edge length. The shear angle and the average cutting edge length are calculated by Eqs. 1.8 and 1.9, respectively.

$$\tan\phi = \frac{r_c \cos\gamma}{(1 - r_c \sin\gamma)} \tag{1.8}$$

where r_c is the chip thickness ratio.

$$l = r_\varepsilon \left[\kappa_r + \arcsin\left(\frac{f}{2r_\varepsilon}\right)\right] + \frac{d - r_\varepsilon[1 - \cos(\kappa_r)]}{\sin(\kappa_r)} \tag{1.9}$$

where r_ε is the tool nose radius, κ_r is the approach angle, f is the feed and d the depth of cut.

Table 1.1 Cutting conditions for particulate MMC

Parameters	Chambers [25]	Davim [21]	Pramanik et al. [19]
Tool material	PCD	PCD	PCD
Nose radius (mm)	1.6	0.8	0.4
Rake angle (deg)	0	0	5
Approach angle (deg)	85	85	90
Cutting speed (m/min)	50–300	250–700	100–800
Feed (mm/rev)	0.2	0.1	0.1–0.25
Depth of cut (mm)	1	1	0.25–1.5
Workpiece	A356-15% SiC	A356-20% SiC-T6	A6061-20% SiC
Yield strength of matrix (MPa)	138	138	110
Average particle diameter (micron)	22.5	20	12

The authors validated the analytical model with their experimental results as well as those published in the literature showing that the theoretical model captured the major mechanisms of machining MMCs and the predicted cutting forces compared well with experimental results. Table 1.1 summarizes the cutting conditions used in the validation of the theoretical model with experimental results published in the literature. The theoretically predicted cutting forces are within 6–11% of the experimental measurements done by Chambers [25] and Davim [21]. Additionally the model developed by Pramanik et al. [19] is compared to the cutting force predictions based on the model by Kishawy et al. [20].

Figure 1.2 compares the predicted and experimental forces with varying feed, where the cutting and thrust forces increase more or less linearly with an increase in the feed. The cutting and thrust forces predicted from the model developed by Pramanik et al. [19] compare better with experimental results as opposed to the predictions made by the Kishawy et al. [20] model. The trend is maintained in the comparison of cutting forces with varying depth of cut (Fig. 1.3). On the other hand the cutting forces decrease approximately linearly with an increase in the cutting speed as shown in Fig. 1.4. Once more the predictions obtained from Pramanik et al. [19] are closer to experimental results than the predictions obtained from Kishawy et al. [20].

Although the theoretical force model predicts cutting forces accurately it still is an approximation of turning into an orthogonal cutting model. This limits the information supplied by the predictive model and hence numerical techniques such as finite element methods have become popular in carrying out machining simulations as will be discussed later in the chapter.

1.4.2 Cutting Temperatures in Machining of MMC

During cutting, high temperatures are generated in the region of the tool cutting edge as a form of cutting energy dissipation. These temperatures have a controlling influence on the rate of wear of the cutting tool and on the friction between the chip and tool, and they can significantly affect the functional performance of a

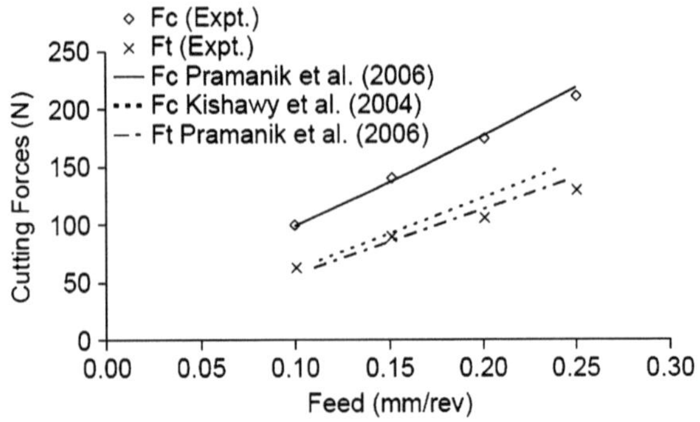

Fig. 1.2 Comparison of predicted and experimental forces with varying feed

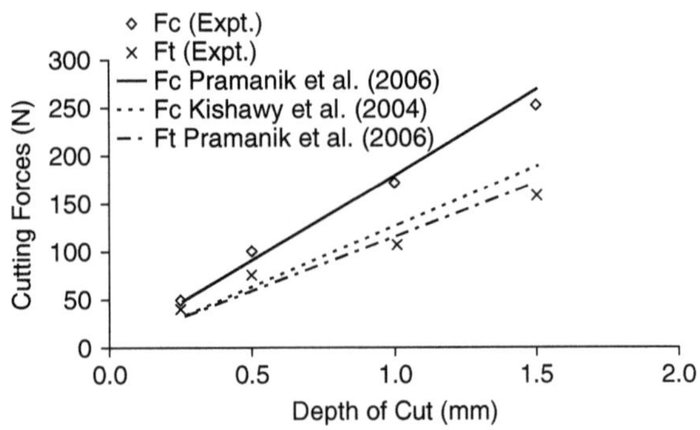

Fig. 1.3 Comparison of predicted and experimental forces with varying depth of cut

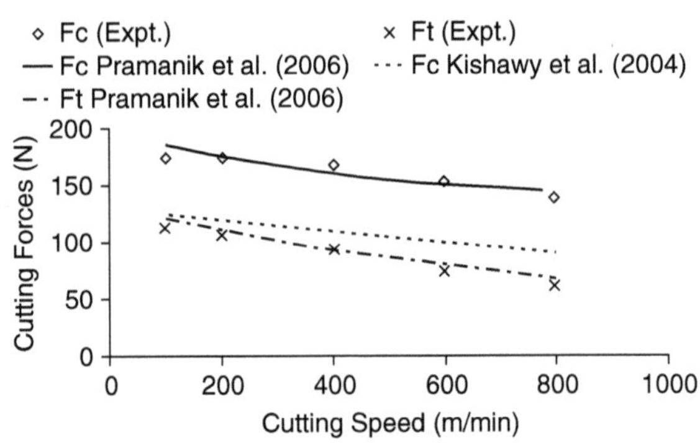

Fig. 1.4 Comparison of predicted and experimental forces with varying cutting speed

machined part due to residual stresses or thermal distortion. Therefore considerable attention has been paid to the measurement and prediction of the temperatures in the tool, chip and workpiece in metal cutting.

Analytical models to exclusively predict cutting temperatures in machining of MMCs do not exist. Therefore many of the studies consider the material to be an equivalent homogenous material and utilize analytical models developed for metal cutting. In metal cutting the material is subjected to extremely high strains and predominantly plastic deformation. By assuming the elastic deformation to only be a very small portion of the total deformation, it is assumed that all the energy required in cutting is converted into heat.

Table 1.2 Values of Γ for machining of an aluminum alloy reinforced by SiC particles

Cutting speed (m/min)	Feed (mm/rev)	Γ
60	0.1	0.52
60	0.3	0.28
180	0.1	0.25
180	0.3	0.17

Workpiece: A359/SiC/20p; tool: CVD coated WC insert; nose radius: 0.8 mm; rake angle: 0°; relief angle: 11°; lead angle: 0°
Source: Liu and Chou [43]

In cutting there are two major regions of plastic deformation responsible for the conversion of energy into heat: the shear zone, or primary deformation zone, and the second deformation zone along the tool–chip contact. There is an additional heat source when the tool has flank wear. This third heat source is a result of the friction between the tool flank face and the workpiece. If the tool is not severely worn the heat generated due to the rubbing of the flank face is negligible and can be neglected. Details on the derivation of the analytical models for calculating the temperatures in metal cutting can be obtained from Boothroyd and Knight [42].

The first heat source is the heat generated due to the shearing of the chip. The energy liberated due to chip shearing (P_s) is given by Eq. 1.10.

$$P_s = F_s v_s \tag{1.10}$$

where F_s is the shear force and v_s is the chip velocity along the shear plane. Using Merchant's theory the shear force is calculated by Eq. 1.11.

$$F_s = F_C \cos\phi - F_T \sin\phi \tag{1.11}$$

where F_C and F_T are the cutting and thrust force and ϕ is the angle of the shear plane given by Eq. 1.8.

A fraction, Γ, of the heat generated in the shear zone is conducted into the workpiece, while the remainder is transported with the chip and is estimated to be approximately 80% of the total heat generated in the shear zone [42]. A similar approximation was used by El-Gallab and Sklad [6] in their study of machining of an aluminum matrix with a 20% by volume fraction reinforcement of SiC particles. Values of Γ for an aluminum alloy (A359) reinforced with 20% by volume fraction of SiC composite were numerically estimated by Liu and Chou [43] and are summarized in Table 1.2. The value of Γ depends on the workpiece material, tool/workpiece combination and the cutting conditions employed; hence care must be taken in using the values shown in Table 1.2.

According to Weiner's relationship [44], the shear plane temperature (θ_s) is evaluated by Eq. 1.12.

$$\theta_s = \frac{(1 - \Gamma)P_s}{(\rho c v_s t_o a_w)} \tag{1.12}$$

where ρc is the volumetric specific heat, t_o is the uncut chip thickness, a_w is the width of cut and Γ is the portion of the heat source conducted back to the workpiece.

Compared with the experimental data obtained by El-Gallab and Sklad [6], the theory provides slightly underestimated results. In the theory a plane heat source was assumed and that heat can only flow into the workpiece by conduction. In reality heat is generated over a wide zone, part of which extends into the workpiece. The effect of this wide heat generation zone becomes increasingly important at high speeds and feeds, therefore the deviation between theoretical and experimental data could be explained.

The second heat source is the heat generated along the tool-chip interface due to friction. The energy liberated due to friction is given by Eq. 1.13.

$$P_f = F_C \mu v \tag{1.13}$$

where F_C is the cutting force which can be calculated using Eq. 1.1, μ is the coefficient of friction and v is the chip velocity which can be given by Eq. 1.14.

$$v = r v_c \tag{1.14}$$

where r is the chip ratio and the v_c is the cutting speed.

The maximum temperature in the chip takes place at the exit from the secondary deformation zone and is given by Eq. 1.15.

$$\theta_{\max} = \theta_o + \theta_s + \theta_m \tag{1.15}$$

where θ_o = initial workpiece temperature, θ_s = temperature increase due to the material passing through the primary deformation zone and θ_m = temperature increase as material passes through the secondary deformation zone. The following solution was obtained by Rapier [45] for θ_m as shown in Eq. 1.16.

$$\theta_m = 1.13\theta_f \sqrt{\frac{R}{l_o}} \tag{1.16}$$

where l_o = ratio of the heat source length to the chip thickness l_f/t_c. R is a dimensionless number given by $\rho c V a/k$ and known as the thermal number, with ρ being the density (kg/m^3), k the thermal conductivity (J/(s(m)K)), c the specific heat capacity (J/(kg K)), V the velocity of the material (m/s) and a the linear dimension which for single point turning is typically the width of cut.

θ_f is the average temperature increase of the chip resulting from the secondary deformation as given by Eq. 1.17.

$$\theta_f = \frac{P_f}{\rho c V t_o a_w} \tag{1.17}$$

A comparison of calculated values of Eq. 1.17 by Rapier [45] with experimental data showed that his theory considerably overestimated θ_m. This could be explained by the fact that the friction-deformation zone, instead of being planar, has a finite width in the analytical model. Once again compared with experimental data obtained by El-Gallab and Sklad [6], the theory overestimates the temperature measurements along the tool rake face. Therefore, from analytical models it is clear that the cutting temperatures deviate from experimental measurements.

Numerical methods such as finite-element models are therefore attractive for studying the cutting temperatures in machining of MMC.

1.4.3 Chip Formation of Particulate MMC Cutting

Chip formation involves the plastic deformation of the shear zone in machining. Traditionally, in metal cutting the shear zone can be analyzed based on the thin-plane model or the thick-plane model. At higher cutting speeds the shear zone approximates to the thin-plane model which allows for easy determination of the shear zone angle. In practice MMCs are machined at relatively high-cutting speeds and therefore the approximation to a thin-plane model is representative of the cutting process. However, the formation of chips cannot be only related to the nature of the shear zone but is also related to the material properties, microstructure and instabilities in the cutting process. The types of chips formed in turning are highly dependent on the cutting conditions. Broadly, the types of chips are classified as (a) Continuous, (b) Built-up Edge, (c) Serrated and (d) Discontinuous and shown schematically and with micrographs in Fig. 1.5.

Typically continuous chips are usually formed when machining at higher cutting speeds and/or machining with cutting tools with a high-rake angle. The deformation of the chip occurs primarily in the primary shear zone marked in Fig. 1.5a with some deformation occurring along the secondary shear zone shown in Fig. 1.5b. Continuous chips are generally not desirable as they tend to get tangled up causing unnecessary delays in machining operations. Chips with built-up edges are generally formed when machining at lower cutting speeds. This phenomenon has been widely observed when machining aluminum matrix MMCs. The built-up edge (Fig. 1.5c) consists of layers of materials from the workpiece that is gradually deposited on the tool. As the material builds up, the BUE becomes larger and subsequently becomes unstable and breaks off. Some of the broken BUE is carried away by the chip while the remaining BUE adheres to the machined workpiece. The BUE is to a certain extent undesirable since it is one of the reasons for a poor surface finish. On the other hand a thin stable BUE is desirable as it is found to protect the tool surface thereby prolonging the tool life. Serrated chips shown in Fig. 1.5d are semi-continuous chips with zones of low-and high-shear strain. The chips demonstrate a sawtooth like pattern and are generally associated with machining of low-thermal conductivity materials like titanium. Discontinuous chips associated with machining of brittle materials, thermosetting polymer composites, MMC at very low- or very high-cutting speeds are shown in Fig. 1.5e. Impurities and hard particles (reinforcements) act as nucleation sites for cracks and therefore result in discontinuous chips.

In machining of MMCs, the most common types of chips are serrated and discontinuous chips with continuous chips formed under certain limited cutting conditions. Lin et al. [3] observed sawtooth-type chips that are most commonly observed while machining titanium alloy. The chip formation mechanism is

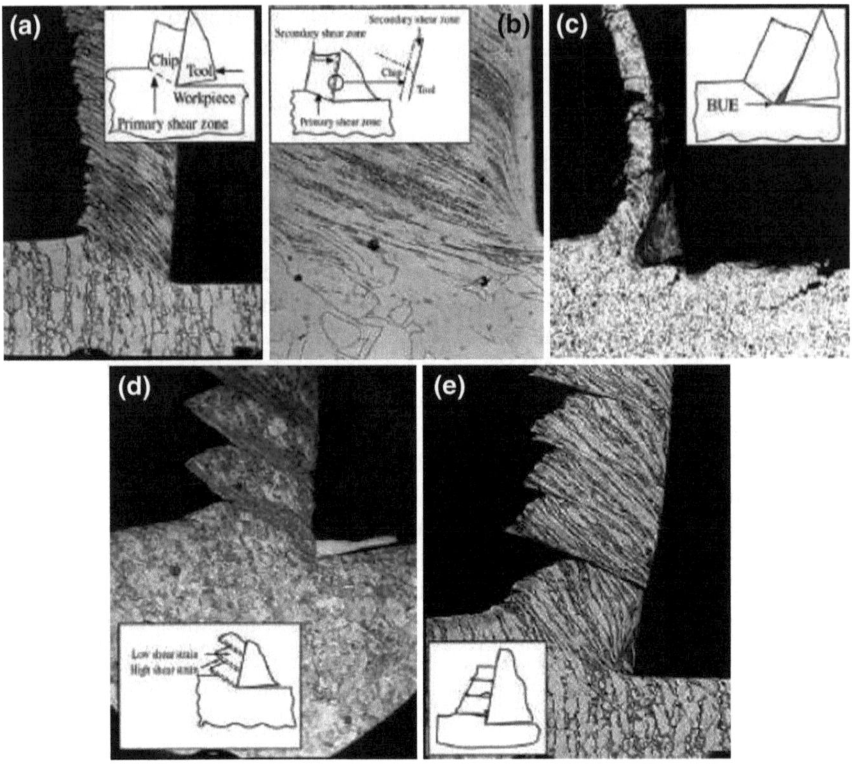

Fig. 1.5 Basic types of chips produced in metal cutting. **a** Continuous chip with narrow, straight primary shear zone. **b** Secondary shear zone at the chip-tool interface. **c** Continuous chip with built-up edge. **d** Segmented or nonhomogenous chip. **e** Discontinuous chip (*source*: S. Kalpakjian [46])

accompanied by a severe plastic deformation of the shear zone. The addition of silicon carbide in the aluminum alloy reduces the ductility of the material contributing to the segmental chip. The chip formation mechanism involves the initiation of cracks, void formation due to the separation of the particles and matrix, subsequent coalescence of the voids, propagation along the shear zone and fracture and sliding of the material to form the semi-continuous chips. Additional tests conducted on aluminum alloy MMCs showed that sharp tools produced continuous chips, while worn tools, higher feed or depth of cut resulted in semi-continuous chips [3, 4, 47]. Researchers have also found similarities in the chip formation mechanism of MMCs to that of conventional monolithic materials. Flow lines associated with particles in the MMCs are similar to the flow lines due to the deformation of grain boundaries of aluminum, titanium and steel [48–50] with the particles aligned along the shear plane in the chip root region. On the other hand the presence of displaced and fractures particles play an important role in machining of MMCs [4, 19]. Therefore the chip formation mechanism is highly

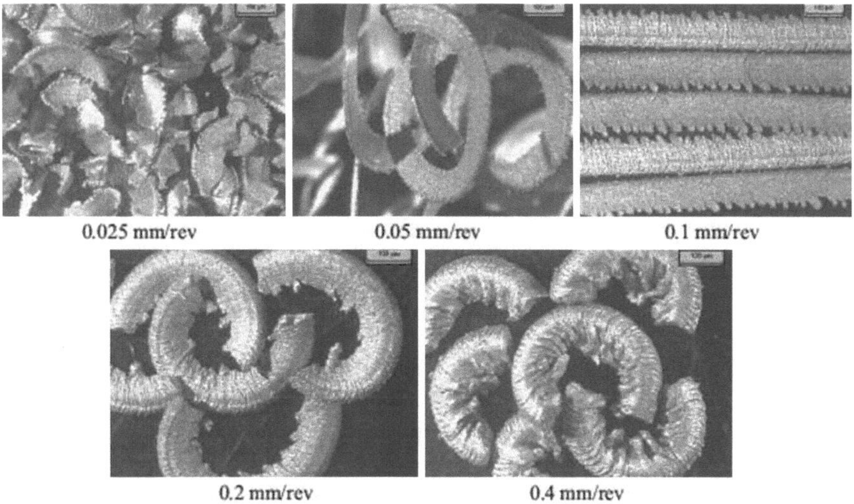

Fig. 1.6 Chip shapes of the MMC at different feeds (at speed of 400 m/min and depth of cut 1 mm) (*source*: Pramanik et al. [10], with permission from Elsevier)

dependent on the particle fracture/displacement, subsequently affecting the cutting forces in machining of an MMC.

Joshi et al. [51] showed the longitudinal micrographs of typical chips formed under machining of aluminum MMCs as a function of particle volume fraction. The cutting was done on a shaper at a low-cutting speed of 16.6 m/min on an aluminum alloy with 0, 10, 20 and 30% in volume fraction of silicon carbide particles. The results indicated that as the volume fraction of the particles increased from 0 to 30% the outer profile of the chips varied from being wavy to a prominent sawtooth-like pattern which was absent at higher cutting speeds. The scanning electron microscope (SEM) images of the chip root were obtained, which clearly show that during the chip formation process fracture is initiated at the outer surface of the chips and propagates toward the tool nose while the rest of the chip is removed by plastic deformation along the shear plane.

In other experimental studies, Pramanik et al. [10] found that the chip breakability improved due to the presence of the reinforcement particles in the MMC. Short chips were formed under all conditions while with the non-reinforced alloy long and unbroken chips of almost the same length were formed under all the cutting conditions. Figure 1.6 illustrates the short and irregular shaped chips formed during machining of the MMC as a function of the feed. At lower feeds of 0.05 and 0.1 mm/rev long spiral and straight chips were observed. With a further increase of the feed (0.2 and 0.4 mm/rev), all the chips became shorter and of a C-shape. On the other hand, for the non-reinforced alloy it was found that in general the chip shape did not change with an increase in the feed or with an increase in the cutting speed (Fig. 1.7). With a variation of cutting speed, very long and brittle chips were formed for the MMC (Fig. 1.8). At lower cutting speeds (100 and

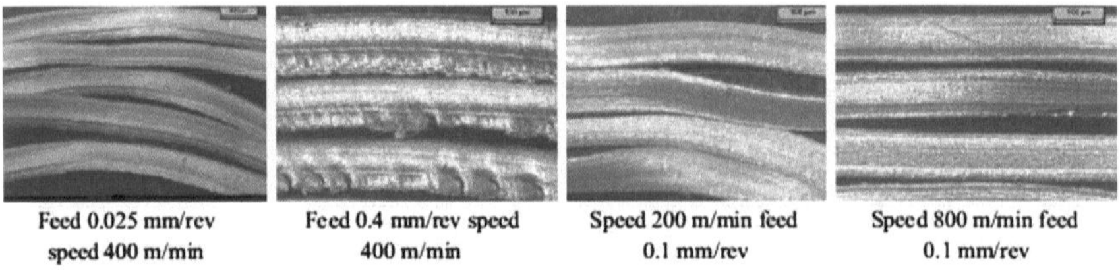

Fig. 1.7 Chip shapes of the un-reinforced alloy at different cutting conditions (depth of cut 1 mm) (*source*: Pramanik et al. [10], with permission from Elsevier)

200 m/min) all the chips were spiral in shape but at higher cutting speeds (400, 600 and 800 m/min) the chips became straight. The sawtooth nature of the chips is also clearly observable for the MMC chips as opposed to the un-reinforced alloy.

1.4.4 Cutting Forces in Fiber Reinforced MMCs

Machining studies of whisker or fiber reinforced MMC are extremely limited. Machining studies conducted on fiber reinforced polymer (FRP) composites provide a starting point in understanding the cutting mechanics of fiber reinforced MMCs. The study done by Koplev et al. [52] is considered as one of the first real attempts at understanding the machining behavior of fiber reinforced composites. They conducted orthogonal machining tests on carbon fiber reinforced polymer (CFRP) composites and observed the chip formation, surface quality and the cutting forces for two fiber orientations: perpendicular (90°) and parallel (0°) fiber orientation relative to the cutting direction. Two important results were observed: the chip formation mechanism was a series of fractures observed in the fibers and a rougher surface was observed for 90° fiber orientation samples as compared to 0° fiber orientation. In another study Takeyama and Iijma [53] described the chip formation process in machining of a glass fiber reinforced polymeric (GFRP) composite. They observed that the chip formation is highly dependent on the fiber orientation with respect to the cutting direction and observed metal-like chip formations while machining the composite with a thermoplastic matrix as opposed to a thermosetting resin polymer matrix. Kim et al. [54] conducted orthogonal tool wear tests on CFRP specimens. Fiber orientation angle and cutting speed were the major contributors to the flank wear, which was the major wear phenomenon observed. The tool wear was caused due to the very abrasive nature of the carbon fiber. It was also shown that the fiber orientation and the feed affected the surface roughness more than the cutting speed. Nayak and Bhatnagar [55] showed that the cutting force and the sub-surface damage increased with increasing fiber orientation while the rake angle had no or minimal effect on the cutting forces and the observed damage. The parameters that are the major contributors to the cutting forces, surface quality and tool wear are fiber orientation, tool geometry and

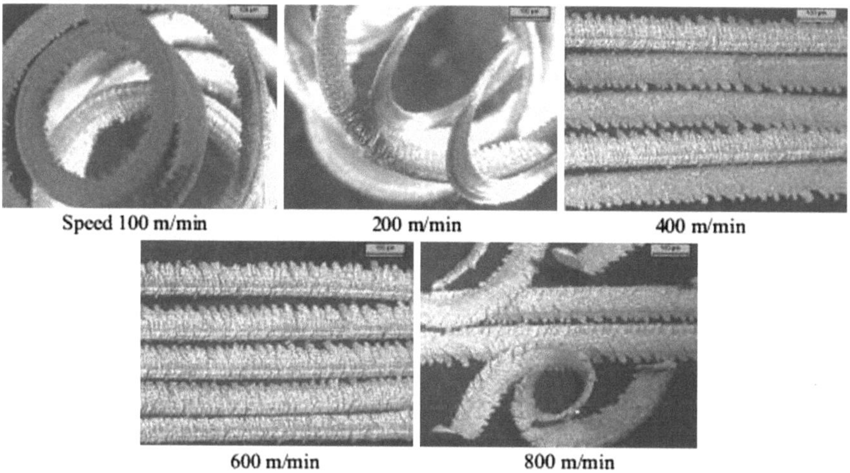

Fig. 1.8 Chip shapes of the MMC at different speeds (at feed 0.1 mm/rev and depth of cut 1 mm) (*source*: Pramanik et al. [10], with permission from Elsevier)

machining parameters. Therefore it is clear that the fiber orientation angle affects the cutting mechanics of machining fiber reinforced composites. It is postulated that the fiber reinforced MMC behaves in a manner similar to FRPs, wherein the fiber orientation plays an important role in determining the cutting mechanism.

Experimentally, Wang and Zhang [56] and Zhang et al. [57] have identified that the fiber orientation relative to the cutting direction θ is the critical parameter in determining the cutting forces and the surface integrity of the machined part. Figure 1.9 schematically shows the cutting of a long-fiber reinforced composite. The model developed here works for fiber orientations between 0 and 90°. In machining of polymer composites there are three regions of interest which are required for prediction of the cutting forces and these are shown in Fig. 1.9. In the first region, fracture occurs at the cross-section of the fibers and along the fiber-matrix interface. This region is in front of the tool rake face with the shear plane in the form of a zig-zag pattern, a result of the fibers being perpendicular to the cutting direction. In the second region, deformation takes place along the nose radius of the cutting tool where the material is pushed down (plowing). In the third region the material which is pushed down bounces back after cutting and contributes to the deformation of a long-fiber reinforced composite. Chips formed in machining of long-fiber or whisker MMCs are invariably discontinuous short chips similar to machining of thermoset polymer composites [58]. In cases where the fiber orientation angle is greater than 90°, the analysis is complicated and can only be accurately resolved numerically.

The total cutting force is a combination of the resultant deformations in the three regions identified in Fig. 1.9. In Fig. 1.9, the first region has a depth a_c and the second region equals the edge radius of the tool (r_e). The positive directions of the cutting forces are taken in the positive y and z directions as shown in Fig. 1.9.

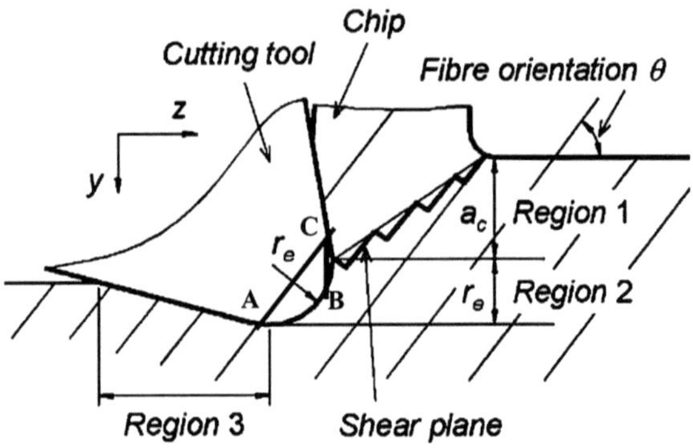

Fig. 1.9 Deformation zones when cutting long-fiber reinforced composites (*source*: Zhang [41], with permission from Elsevier)

Therefore, the total cutting force in the vertical and horizontal directions is given by F_y and F_z by Eqs. 1.18 and 1.19, respectively.

$$F_y = F_{y1} + F_{y2} + F_{y3} \tag{1.18}$$

$$F_z = F_{z1} + F_{z2} + F_{z3} \tag{1.19}$$

where F_{yi} and F_{zi} ($i = 1, 2, 3$) are the corresponding forces in the three regions and are given by Eqs. 1.20–1.25.

$$F_{z1} = \tau_1 h a_c \frac{\sin\phi \tan(\phi + \beta + \gamma_o) + \cos\phi}{\left({}^{\tau_1}/_{\tau_2}\right)\cos(\theta - \phi)\sin\theta - \sin(\theta - \phi)\cos\theta} \tag{1.20}$$

$$F_{y1} = \tau_1 h a_c \frac{\cos\phi \tan(\phi + \beta + \gamma_o) - \sin\phi}{\left({}^{\tau_1}/_{\tau_2}\right)\cos(\theta - \phi)\sin\theta - \sin(\theta - \phi)\cos\theta} \tag{1.21}$$

$$F_{z2} = P(\sin\theta + \mu\cos\theta) \tag{1.22}$$

$$F_{y2} = P(\cos\theta + \mu\sin\theta) \tag{1.23}$$

$$F_{z3} = \frac{1}{2} r_e E_3 h \cos^2\gamma_o \tag{1.24}$$

$$F_{y3} = \frac{1}{2} r_e E_3 h (1 - \mu\cos\gamma_o \sin\gamma_o) \tag{1.25}$$

where $\phi \approx \arctan\left[\frac{\cos\gamma_o}{1-\sin\gamma_o}\right]$, E_3 is the effective modulus of the composite, h is the thickness of the workpiece, v is the Poisson's ratio, P is the resultant force and empirically given by $\frac{1}{2}\arctan\left(\frac{30}{\theta}\right)$ for the polymer composite, μ is the coefficient of friction, τ_1 and τ_2 are the shear strengths of the material in the AC and BC directions which are marked on Fig. 1.9, γ_o is the tool rake angle and β is the friction angle.

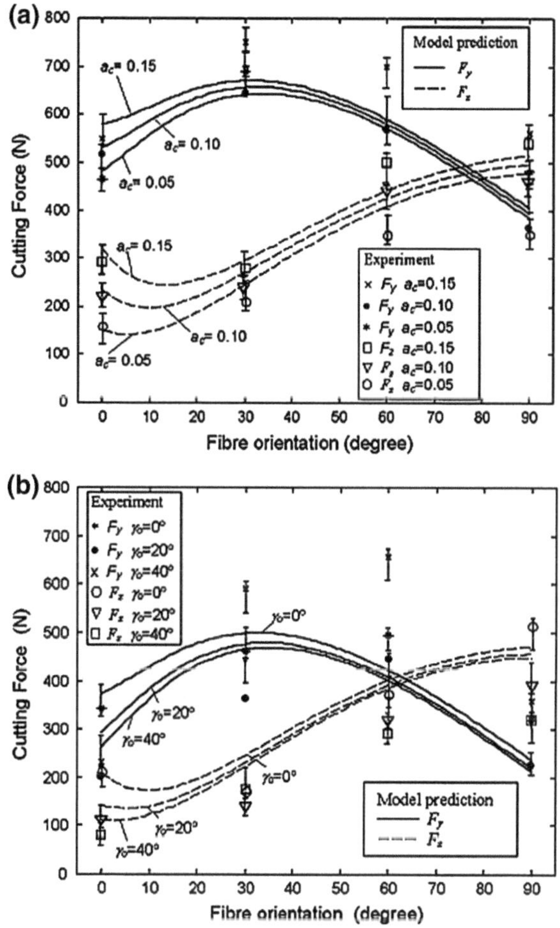

Fig. 1.10 Comparison of the model predictions vs. experimental measurements for the cutting forces for two materials. **a** Variation with depth of cut and fiber orientation for material with $E_3 = 5.5$ GPa. **b** Variation with the tool rake angle and fiber orientation for material with $E_3 = 3.5$ GPa (*source*: Zhang [41], with permission from Elsevier)

Validation of the theoretical model was done by comparing the predictions with experimental results for two long-fiber polymer composites by Wang and Zhang [56]. The authors experimentally determined the model parameters $\tau_1 = 90$ MPa, $\tau_2 = 20$ MPa, $\beta = 30°$, $\mu = 0.15$, and $v = 0.026$. Figure 1.10 shows the comparison between the model predictions and the experimental measurements for fiber angles in the range of 0–90° where the trend in the cutting force variation is well represented with the model being able to capture most of the deformation mechanisms.

Due to the limited experimental data available for machining of long-fiber MMCs, predictions for machining a MMC with $\theta = 90°$ are shown here. The workpiece material, cutting conditions and the model parameters are summarized in Table 1.3. The model predicts the cutting force (F_c) to be 53 N, while experimentally the cutting force was measured to be 49 ± 3 N. Similarly, the thrust force (F_t) is predicted to be 51.4 N which compares reasonably well with the experimental measurement of 57 ± 4 N. Experimentally, the thrust force was

Table 1.3 Cutting conditions for machining a long-fiber MMC (Dandekar and Shin [36])

Workpiece	Al-2%Cu matrix—60% by volume fraction of alumina fibers		
Tool material	PCD	E_3 (GPa)	160
Nose radius (mm)	0.8	μ	1.14
Rake angle (deg)	5	v	0.27
Cutting speed (m/min)	30	$\tau_1 = \tau_2$ (MPa)	176
Feed (mm/rev)	0.02	$\phi(°)$	47.5
Depth of cut (mm)	0.5	r_e (μm)	20
P(N)	28.5	β (°)	48.7

consistently higher than the cutting force which is not predicted by the model. It is understandable that the model will not be able to capture the entire deformation phenomena during machining of MMCs, especially since the model was developed for FRP machining. Nonetheless it provides a reasonable estimate of cutting forces and also reiterates the importance of numerical modeling.

1.5 Finite-Element Modeling in Machining

A tremendous amount of research has been done in understanding the mechanics associated with machining processes. Compared to empirical or analytical methods, the development of computers has allowed researchers to study machining through sophisticated numerical techniques. These include finite-element modeling, molecular dynamics studies and recently multi-scale modeling. In this chapter only the former modeling technique is addressed. Despite the success in modeling 2D and 3D machining, there are still many challenges associated with modeling of machining. A bibliography of all the finite-element modeling research done from 1976 to 2003 for machining was compiled by Mackerle [59, 60]. Over 300 papers have been referenced in this bibliography. In spite of this, research in this field continues to grow. Active research continues to provide an understanding of the constitutive behavior of materials, chip formation, modeling of composite machining, tool–chip interface behavior, etc. The approaches that have been used for numerical modeling of machining processes range from atomistic level techniques to macro level continuum mechanics methods. Despite these efforts problems have not yet been completely solved due to the extreme complexity involved in machining processes.

1.5.1 FEM Formulations and Approaches

In FEM, three main formulations have been proposed for machining simulations: Eulerian, Lagrangian and Arbitrary Lagrangian Eulerian (ALE) methods. In the Eulerian method, mesh is spatially fixed while material is allowed to flow through the meshed control volume. The advantage of the Eulerian method is that the excessive element distortion is absent since the mesh is fixed. The disadvantage is that the initial shape of the chip and the contact conditions are needed to be known in advance and hence it is not suitable for machining simulations. In the Lagrangian approach on the other hand the mesh is attached to the workpiece and the elements are allowed to deform similarly to machining. This method has been the most popular method in machining simulation as the chip geometry does not have to predefined, but rather develops as the cutting progresses. However it is dependent on a defined chip-separation criterion [61]. The disadvantage of the method is the excessive element distortion and the need for frequent remeshing that is computationally expensive. The ALE is an adaptive meshing technique that combines pure Lagrangian and Eulerian formulations to incorporate the advantages of both. In the ALE method, the finite element mesh is neither fixed spatially nor attached to the material. Instead it is allowed to flow with the material [62]. In this manner severe distortion of the elements is avoided without the need for remeshing. The ALE method has successfully been implemented in machining simulation to predict chip formation in metal cutting by numerous authors [63–70]. A good review on FEM of metal cutting was provided by Soo and Aspinwall [61] for the interested reader.

Numerous numerical modeling studies have been conducted on orthogonal machining of composite materials. Three primary approaches have been successfully implemented: (a) a micromechanics-based approach (b) an equivalent homogeneous material (EHM) based approach and (c) a combination of the two approaches. The micromechanics and the equivalent homogenous material (EHM) based approaches have their respective advantages and disadvantages [71]. The micromechanics approach describes the material behavior locally, and hence it is possible to study local defects such as debonding and complicated deformation mechanisms especially in fiber reinforced composites. The required computation time however, is very high since to predict local damage the mesh used for this study is a lot finer than the one needed for the EHM model. On the other hand the EHM approach reduces the computation time but is not capable of predicting the local effects, namely, the damage observed at the fiber-matrix interface [72–74]. Therefore there is a need to harness the advantages of both the continuum and micromechanics models in their capabilities of predicting cutting forces and sub-surface damage. Rao et al. [75, 76] used a combination of the EHM and micromechanical model to model 2D orthogonal cutting and study the effect of fiber orientation on the simulated cutting forces, chip formation, the extent of fiber damage, matrix damage and debonding. The micromechanical model was used in the vicinity of the tool, while the EHM model was implemented away from the tool.

1.5.2 Constitutive Materials Modeling

In any FEM model an essential input is the accurate definition of material properties. Under machining conditions, generally the workpiece is subjected to extreme deformation involving high levels of strain and strain rates and rapid temperature rise. The workpiece material is usually modeled by constitutive equations describing the stress-strain response together with its dependence on strain rate, temperature and work hardening. Furthermore in the Lagrangian or ALE analysis there is a necessity to include a chip-separation criterion. This chip-separation criterion triggers the material fracture resulting in separation of a chip from the workpiece. To achieve this one must be able to accurately describe the fracture behavior to represent the material under study through an appropriate damage model. Any constitutive model selected for the material needs to be validated prior to inputting into a finite-element model.

In machining simulations when using an EHM approach for the machining of the composite, homogenization of the material is necessary. Traditional homogenization techniques based on either exact or energy methods are useful primarily in the elastic regime of the composite. In machining a high degree of plasticity is involved in the primary and secondary shear zones, and extremely high-strain rates and temperatures are also observed. Therefore one also needs a description of the plastic deformation of the composite material and also its behavior at high temperatures and high strain-rates. To this end a number of empirical constitutive models have been presented for modeling the deformation behavior at high strain-rates and temperatures: for instance, the widely used Johnson–Cook (J–C) equation [77] and the Norton-Hoff law [78, 79]. The Norton-Hoff material model was applied by Monaghan and Brazil [22] to model the machining of a particulate reinforced MMC while the J–C model has been successfully applied to the modeling of a silicon carbide particle reinforced aluminum matrix composite [80, 81]. The Johnson–Cook equation is based on experimentally determined flow stress as a function of strain, temperature and strain rate in separate multiplicative terms. This equation therefore does not consider the interactions between the terms. The model is relatively easy to apply in the FEM setting and hence has been used in many studies. The important aspect in applying this model is the relevance of the experimental data as it needs to cover material deformation under a range of strains, strain rates and temperatures typically seen in machining and hence it is expected to obtain data from a number of resources. One caveat in using empirical models is that its choice significantly influences the predicted results [61] and so it is extremely critical to choose the right material model and parameters.

On the other hand for carrying out multi-phase modeling, good material models for the reinforcement and matrix along with the interface are necessary inputs. Once again dependent on the reinforcement or the matrix it is necessary to select an appropriate material model. A number constitutive models are available in the literature to model the matrix material; Zerrili–Armstrong (Z–A) type constitutive model [82], a Johnson–Cook type model [77], Mecking–Kocks model [83] and a physics-based model proposed by Nemat-Nasser et al. [84] for OFHC copper etc. Another material model successfully applied to modeling machining of an aluminum MMC is the Cowper–Symonds (C–S) model for the 6061 aluminum matrix [19]. Any model that is able to capture the high adiabatic shearing observed during machining of MMCs is suitable for simulations. Since both categories, phenomenological and physics-based model, determine material constants by fitting to stress-strain responses at different temperatures and strain rates, the advantage in using one model over another is merely dependent on the number of constants to determine. The Meckings–Kocks model has 23 constants, 12–8 constants are necessary to use the Nemat-Nasser and Li model depending on the inclusion of dynamic strain aging or not, Z-A model needs seven constants but modification to the Z–A model can reduce the number of constants to six [82], the J–C model needs only 5 constants while the C–S model has two fitting parameters. Most of these models have the ability to model the material response as close to experimental data and hence there is always an advantage in using a material model with fewer constants. An advantage of using the Z–A or J–C model is that a number of studies have been conducted to calculate the material constants for aluminum, steel, titanium etc., which are widely available.

The brittle reinforcement materials are usually modeled as linearly elastic materials with damage initiation being instantaneous and hence do not need a damage evolution law in machining simulations. Material properties for a popular reinforcement material of alumina and silicon carbide particles are provided in Tables 1.4 and 1.5. Once again the material properties can be obtained from experimental readings or in some cases through atomistic simulations carried out with techniques such as molecular dynamics. Material failure in the reinforcement is related to the critical stress to the damage equivalent stress at failure. The material is considered failed once the critical stress equals the ultimate tensile stress of the material [85]. Another method for brittle materials was applied by Marusich and Ortiz [86], who applied a failure criterion based on the effective plastic strain of ductile failure. The failure occurred when the stress attained a critical value as determined by the fracture toughness of the material.

The interface between the reinforcement phase and matrix phase plays a crucial role in the analysis of damage or failure in a composite material. Modeling of the interface in a composite structure is achieved through the use of interface elements. A number of different approaches have been considered for interface elements [90] but in general the interface elements function in a similar manner, wherein they connect the two phases while transferring the traction between them until a predefined displacement criterion is reached, at which point the interface element degrades in material stiffness, hence carrying no load. The functional relationship is defined by a traction-displacement relationship. Cohesive elements, a type of interface elements, are widely used to model delaminations and debonding in composite structures as it considers both damage and fracture mechanics and have been successfully applied to composite machining simulations [36, 55, 74, 91]. The interested reader in understanding the application and numerical implementation of the cohesive zone model is encouraged to review the work done by Needleman [92], Tvergaard [93], Xu and Needleman [94], Camacho and Ortiz [95] and Chandra et al. [96] who have studied the application of the cohesive zone model to simulate the metal-ceramic interfaces.

Table 1.4 Mechanical properties of alumina fiber (Nextel™ 610) [87, 88]

Diameter (μm)	14–20
Modulus of elasticity (GPa)	373
Tensile strength in fiber direction (S11) (MPa)	3,100
Compressive strength in fiber direction (S11) (MPa)	3,360
Tensile strength in transverse direction (S22) (MPa)	318
Compressive strength in transverse direction (S22) (MPa)	362
Density (ρ) (kg/m^3)	3,900
Poisson's ratio	0.27
Melting temperature (°C)	2,000
Coefficient of thermal expansion/°C (range: 100–1,100°C)	8×10^{-6}

Table 1.5 Properties of silicon carbide [89]

K	Fracture toughness	3.9 MPa
E	Youngs modulus	408 GPa
ν	Poisson's ratio	0.183
ρ	Density	3.2 gm/cm^3
α	Coefficient of linear thermal expansion	5.12×10^{-6}/°C

1.5.3 Chip-Separation Criteria

Material separation is a complex phenomenon involving many physical processes occurring at the micromechanical level. Fracture begins at the micromechanical scale and eventually with damage accumulation macroscopic fracture is observed. In FEM simulations, the variables that control fracture are the current variables of stress and strain tensors and their histories. Damage initiation is dependent on the current stress/strain state while damage evolution is dependent on the history of the stress/strain state. Some of the damage criteria used today, either built-in in commercial FEA codes or through the implementation of user-defined subroutines for machining simulations, are: (a) constant equivalent strain criterion, (b) maximum shear stress criterion, (c) Johnson–Cook fracture model and (d) Cockroft–Latham criterion. The equivalent strain criterion has been a popular failure criterion for metal cutting simulations [19, 97–99]. In this approach fracture is assumed to occur at the material nodal points when the equivalent plastic strain reaches a critical value dependent on the material. On reaching the critical value the node in front of the tool tip is separated from the workpiece, resulting in chip formation. The drawback of this method is the node separation technique that is computationally intensive and tedious. Similarly a critical stress criterion has also been suggested where node separation is activated once the material reaches a critical stress value [100]. The Johnson–Cook failure criterion is based on the postulation that the critical equivalent fracture strain is a function of the stress triaxiality, strain rate and temperature. The fracture model is semi-empirical in nature and necessitates the determination of constants from tensile tests with high triaxility, shear tests and Hopkinson bar torsion tests at varying temperatures and strain rates, and has been used to model machining of a particulate MMC [80]. This limits the use of the model as a number of experiments are needed to arrive at the five material constants necessary in implementing the model. Another fracture model implemented in machining codes is the Cockcroft–Latham fracture criterion [101, 102]. This criterion is widely used in machining simulations, such as predicting serrated chip formations as is the case with titanium machining. The criterion was developed for bulk forming operations and therefore there is a limitation on its application, as it is applicable only in small and negative triaxiality situations. Fracture in this model occurs when the equivalent strain modified by the principal tensile stress reaches a predefined critical value.

1.5.4 Friction at the Tool–Chip Interface

Another parameter important to the accuracy of numerical machining simulations is the influence of the friction conditions at the tool–chip interface. Friction at the tool–chip interface is a very complex process. One of the methods is to experimentally obtain the coefficient of friction and apply it as a constant over the tool–chip contact length. In most studies, the Coulomb friction model is commonly used as in machining simulations. In their machining simulations of a 20% by volume fraction of particulate (SiC) in an aluminum matrix, Pramanik et al. [19] used a Coulomb friction law and Tresca shear stress limit to model the sticking and sliding conditions at the tool–chip interface. In their study, the friction at the tool–chip interface is controlled by a Coulomb limited Tresca law which is expressed by Eqs. 1.26 and 1.27.

$$\tau = \mu\sigma_n \tag{1.26}$$

$$|\tau| \leq \tau_{\lim} \tag{1.27}$$

where $\tau_{\lim}$ is the limiting shear stress, τ is the equivalent shear stress, μ is the friction coefficient and σ_n the normal stress (contact pressure). According to this model the two contacting surfaces carry shear stresses up to a certain magnitude prior to starting sliding relative to each other. When $\tau > \tau_{\lim}$ the two surfaces slide relative to each other. The limiting shear stress was 202 MPa and the coefficient of friction was modeled as 0.62 [19].

Recently, Filice et al. [103] analyzed the influence of different friction models on the results of numerical machining simulations. They concluded that for the studied workpiece/tool couple, most mechanical results are not influenced by the friction model except the temperature at the tool–chip interface. Nonetheless the accurate representation of the coefficient of friction allows for accurate prediction of cutting forces and temperature distributions.

1.6 Modeling of MMC Machining

It is clear that the presence of reinforcement makes MMCs different from monolithic materials due to incorporation of its superior physical properties. Different failure mechanisms play an important role during the machining of MMCs: (1) cracking of the reinforcing particles; (2) partial debonding at the particle/matrix interface resulting in the nucleation of voids and (3) the growth and coalescence of voids in the matrix. From the previous section on analytical modeling of MMCs machining it is understood that these models are not able to adequately handle all the aforementioned failure phenomena which contribute to the machining performance of the MMC. To handle the tool–reinforcement interaction, finite element simulations of machining have been used. Finite element simulations have been able to successfully predict phenomena such as: (1) flow of particles in the chip root region, (2) debonding of the particles in the secondary cutting zone and sub-surface, (3) tool–workpiece (particle) interaction, (4) pull-out of the particles and (5) fiber debonding and fiber damage in a fiber reinforced MMC.

1.6.1 Modeling Machining of Particulate reinforced MMC

A number of attempts have been made in modeling machining of MMCs [19, 23, 36, 48, 80, 104]. Except for the study done by Dandekar and Shin [80] all the other studies have primarily focused on 2D modeling of orthogonal cutting, which is not realistic for actual machining. Monaghnan and Brazil [48] studied the failure at the particle-matrix interface and the residual stress in machining MMCs using the 2D finite element code FORGE2 for an A356 aluminum alloy reinforced with 30% by volume fraction of silicon carbide, but tool–particle interaction was neglected. El-Gallab and Sklad [23] simulated the residual stresses and sub-surface damage observed in a SiC particle reinforced aluminum alloy and concluded that the feed had the largest effect on the sub-surface damage and the simulated residual stresses, wherein damage and residual stress increased with an increase in the feed. This study, focused on predicting sub-surface damage, lacked in their representation of the interface, since the particles are considered to be perfectly bonded to the matrix, and was concerned with 2D orthogonal machining. Later, others studied the tool–particle interaction by considering particles along, above and below the cutting path [19, 104] as shown in Fig. 1.11.

This section further provides an example for setting up an FEM for modeling machining of particulate MMCs. For the orthogonal machining simulation of particulate reinforced MMC a combination of 2D plane strain continuum quadrilateral (CPE4R) and triangular (CPE3) elements are used for meshing the particles and matrix, while the interface layer is modeled using the 2D cohesive elements. A plain strain analysis is generally used in modeling orthogonal

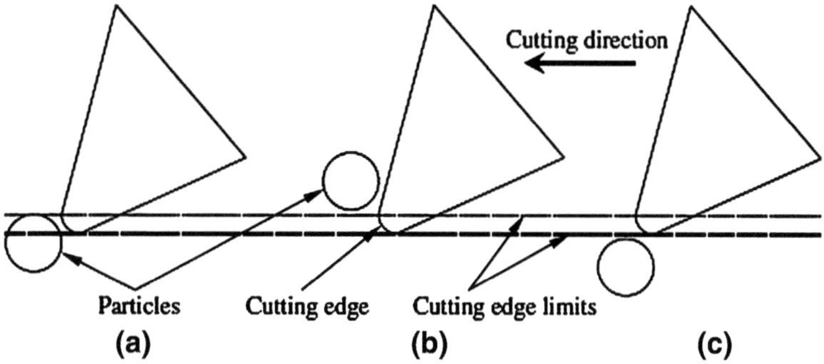

Fig. 1.11 Particle locations with respect to the cutting path: particles **a** along, **b** above and **c** below the cutting path (*source*: Pramanik et al. [19], with permission from Elsevier)

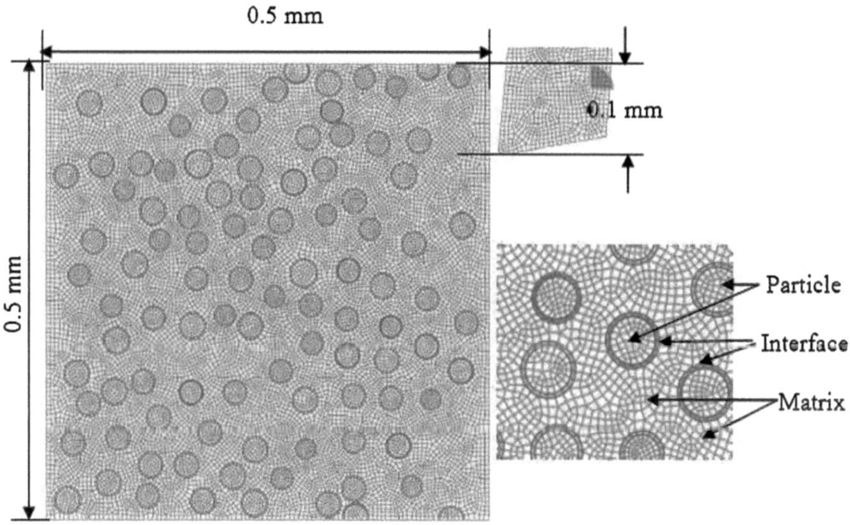

Fig. 1.12 **a** FEM mesh for modeling of machining a particulate MMC showing the length scales. **b** Close-up of the mesh

machining for homogenous materials. The MMC due to the random distribution of particles behaves in a homogenous manner. The random distribution of the particles is carried out using a random particle distribution scheme which is explained later. A representative mesh for the particulate reinforced MMC is shown in Fig. 1.12. For the simulation the boundary conditions applied are as follows; the workpiece is constrained to move in both the (x) and (y) directions at the bottom side, left-hand side and the lower right-hand side. The tool is given a constant velocity in the (x) direction and the tool movement in the (y) direction is constrained. The tool material simulated is a PCD tool with a nose radius of 0.4 mm

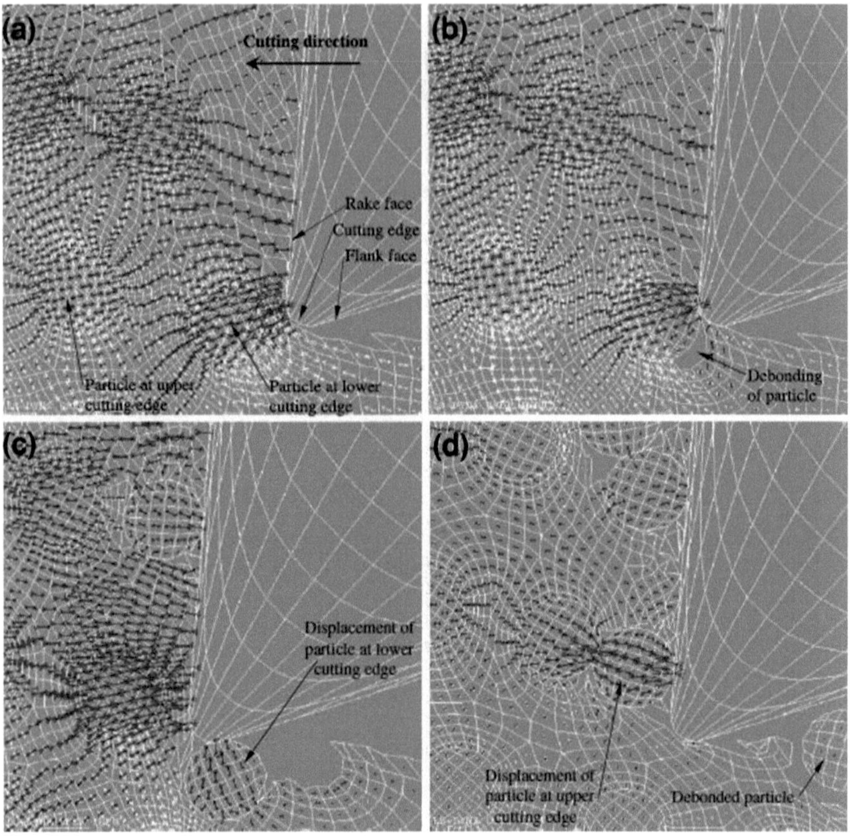

Fig. 1.13 Evolution of stress fields for particles along the cutting path. Compressive and tensile stresses are represented by *black* >-< and *white* <-> symbols, respectively. Their lengths represent the comparative magnitudes (*source*: Pramanik et al. [19], with permission from Elsevier)

and a rake angle of 5°. The cutting conditions are as follows; feed rate of 0.1 mm/rev, depth of cut of 0.25 mm and a cutting speed of 200 m/min.

1.6.1.1 Tool–Particle Interaction

Pramanik et al. [19] conducted an extensive study to characterize the evolution of the stress field and the development of the plastic zone in orthogonal machining of a 20% by volume fraction SiC in a 6061 aluminum matrix. The authors divided the scenarios into three categories: (1) particles along the cutting path, (2) particles above the cutting path and (3) particles below the cutting path.

In the first scenario, the particle is along the cutting path and interacts between the upper and lower limits of the cutting edge as shown in Fig. 1.13a. The evolution of the stress fields during machining in this scenario is captured in

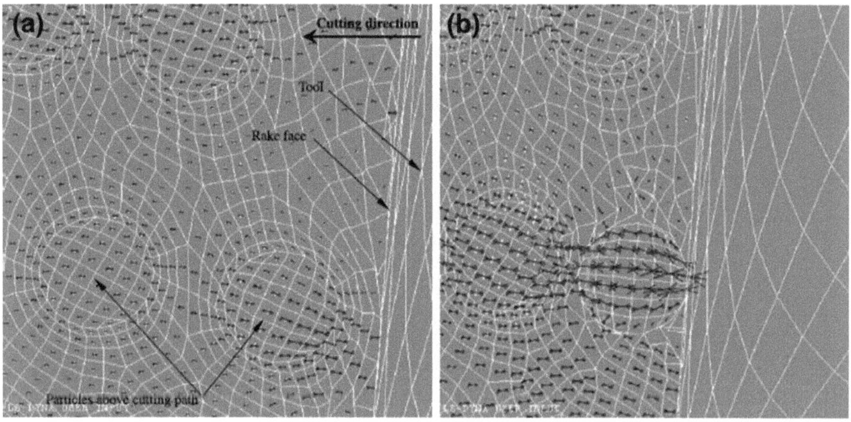

Fig. 1.14 Evolution of stress fields for particles above the cutting path. Compressive and tensile stresses are represented by *black* >-< and *white* <-> symbols respectively. Their lengths represent the comparative magnitudes (*source*: Pramanik et al. [19], with permission from Elsevier)

Fig. 1.13a–d. For a particle located in the lower part of the cutting edge, compressive and tensile stresses are perpendicular and parallel in front of the cutting edge causing possibly fracture of the particles and debonding at the interface. In the simulations particle fracture was not observed as the material definition of the SiC particles lacked any failure criteria. Upon advancement of the tool, the matrix along the upper part of the particle and tool becomes highly compressive while the lower-right interface of the particle becomes tensile (Fig. 1.13a), resulting in the debonding of the particle with the advancement of the tool. During tool–particle interaction significant tensile and compressive stresses are found to be in the left part of the particle (Fig. 1.13b) with the right corner experiencing compressive stress. Upon further advancement of the cutting tool, the particle debonds and plows through the matrix creating a void (Fig. 1.13c) and then slides along the flank face of the tool (Fig. 1.13d). The particle located at the upper part of the cutting edge moves slightly upward due to the plastic flow of the matrix. The stress distribution in this state also promotes particle debonding and/or fracture. Upon further advancement of the cutting tool the two particles interact with each other and are consequently under highly compressive stresses which may cause fracture of the particle as well as contribute to the wear of the flank face.

In the second scenario for particles above the cutting path, the evolution of the stress fields is shown in Fig. 1.14a and b. At the start of machining highly compressive stresses perpendicular to the tool rake face through the particle and in the matrix between the particle and the rake face are observed. Additionally, parts of the particle and the interface are under compressive and tensile stresses as shown in Fig. 1.14a. The combination of this stress state may initiate particle fracture and debonding. Upon further advancement of the cutting tool, the particle partially debonds and the contact region where the particle interacts with the rake face is

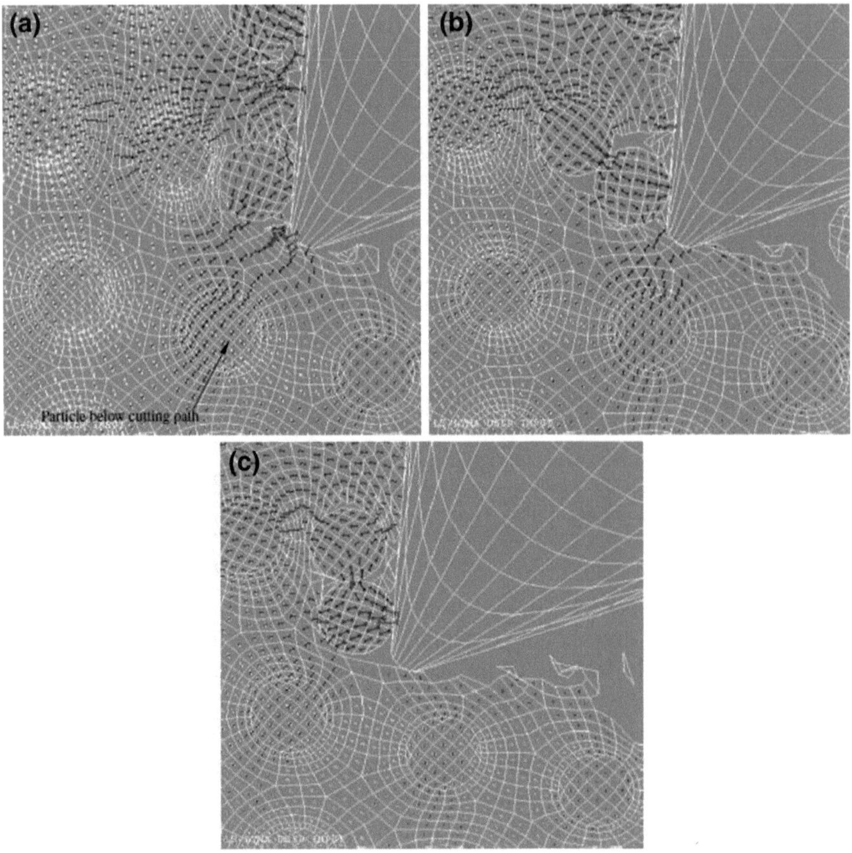

Fig. 1.15 Evolution of stress fields for particles below the cutting path. Compressive and tensile stresses are represented by *black* >-< and *white* <-> symbols respectively. Their lengths represent the comparative magnitudes (*source*: Pramanik et al. [19], with permission from Elsevier)

under high-compressive stresses and encourages particle fracture. Further advancement of the tool, the first particle then interacts with the second particle and slides across the rake face of the tool (Fig. 1.14b).

In the third scenario of the particle below the cutting edge, the stress distribution in the particle and matrix has a direct influence on the residual stress of the machined part. As the tool approaches the particle, the matrix between the cutting edge and particle is under compressive stresses acting in a radial direction to the cutting edge (Fig. 1.15a–c). In this configuration the particle and the matrix are under compressive and tensile stresses which act in the radial direction. Once again this stress distribution promotes particle debonding and void formation in the matrix. When the tool passes over the particle the direction of the tensile stresses becomes parallel to the machined surface while the compressive stresses remain radial to the cutting edge. The newly generated surface (Fig. 1.15c) is

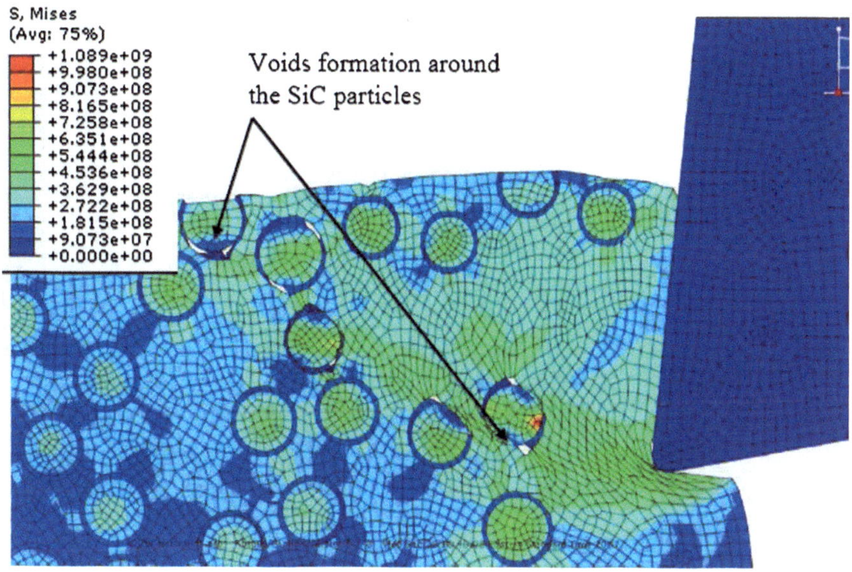

Fig. 1.16 Machining simulation of Al/SiCp MMC showing the failure of the interface

under compressive residual stresses which are parallel to the machined surface. Similar observations were reported by Quan and Ye [105] in their study on machining of a SiC particulate reinforced MMC.

1.6.1.2 Chip Formation and Debonding

The physical deformation phenomenon during machining of MMC is captured well in the model described in Dandekar and Shin [80]. There is distinct shear localization at the initial debonding sites along the interface appearing as a shear band (Fig. 1.16). The particle distribution has a strong effect on the formation and intensities of shear bands between the particles, as well as on the concentration of stress maxima in the vicinity of the particles. This high-stress/strain region along the interface results from the high difference in the elastic modulus between the matrix and the reinforcing particle.

The simulation results are compared to the experimental observations by El-Gallab et al. [8]. The authors concluded that the failure in the composite initiates along the voids which are formed around the SiC particles. The chip formation was segmented and discontinuous with ductile tearing at the edges due to the alignment of the SiC particles. The voids join up forming micro-cracks and subsequent fracture along the shear band. All these observations are clearly seen in the simulations, where there is first void formation along the particles, then alignment of the particles and finally fracture along the shear band (Figs. 1.16, 1.17, 1.18, 1.19).

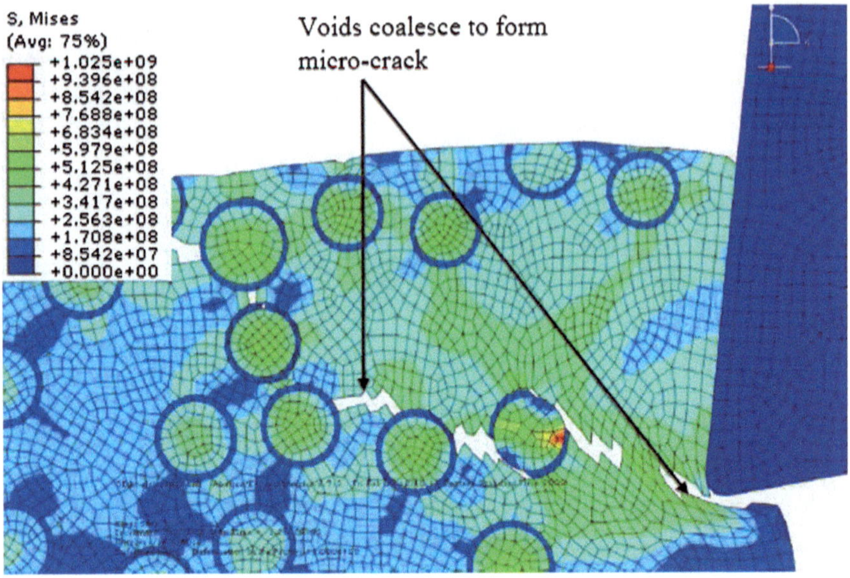

Fig. 1.17 Machining simulation of Al/SiCp MMC showing the void coalescence

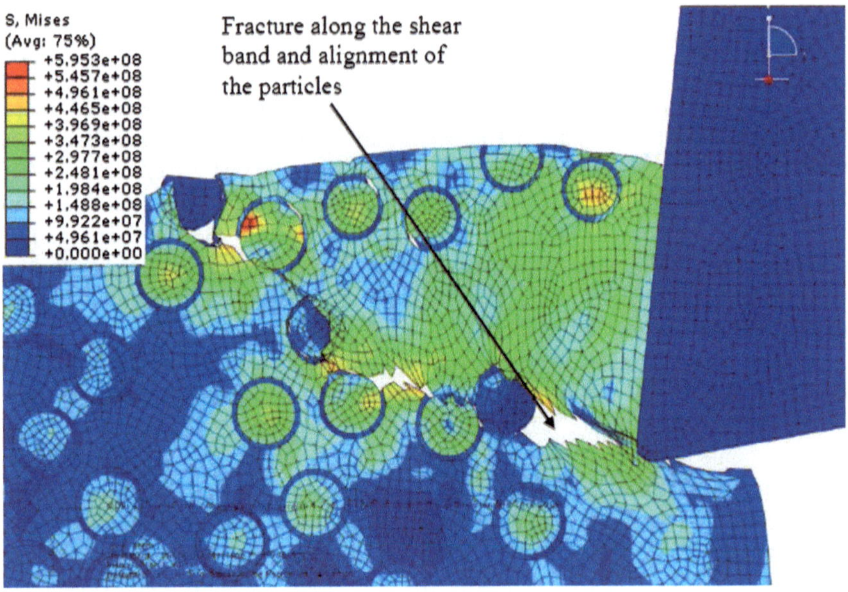

Fig. 1.18 Machining simulation of Al/SiCp MMC showing fracture and alignment of particles

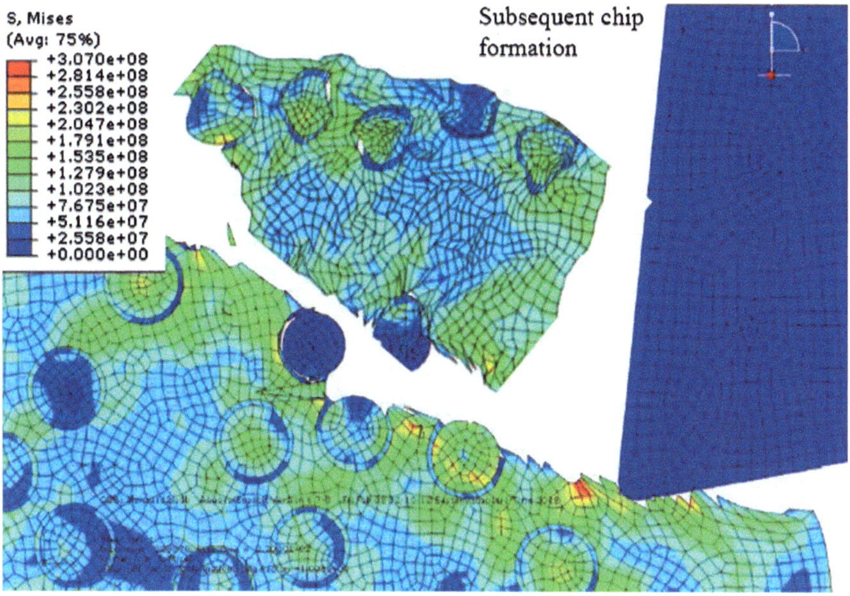

Fig. 1.19 Machining simulation of Al/SiCp MMC subsequent chip formation

1.6.1.3 Numerical Predictions of Cutting Temperature

It was previously shown that theoretical calculations are inadequate in predicting the cutting temperatures in machining. An attractive alternative is to utilize numerical methods such as finite-element models. A number of studies have been conducted in modeling the heat transfer mechanisms involved during the cutting of a metal matrix composite [6, 43, 104, 106, 107].

Among these studies, Zhu and Kishawy [104] simulated the machining of a 10% by volume fraction of Alumina in an aluminum alloy 6061 matrix. The machining model setup corresponds to an orthogonal machining model with particles randomly distributed in the matrix. The two-dimensional orthogonal cutting process was simulated using the commercially available finite element solver ABAQUS/Explicit using the ALE formulation. Two-dimensional four-node displacement and temperature continuum elements featuring reduced integration and hourglass control are used to carry out the coupled temperature-displacement analysis.

The authors identified three main heat sources in the secondary deformation zone: the plastic deformation in the chip which is in contact with the tool rake face, the heat generated due to friction along the tool–chip interface, and the heat generated due to the chip sliding across the rake face. The results indicated that the tool–chip interface friction is the most important factor in the increasing temperature. Figure 1.20a–c shows the contour plots of the temperature for feeds of 0.1, 0.2 and 0.3 mm/rev, respectively. The temperature generated along the tool–

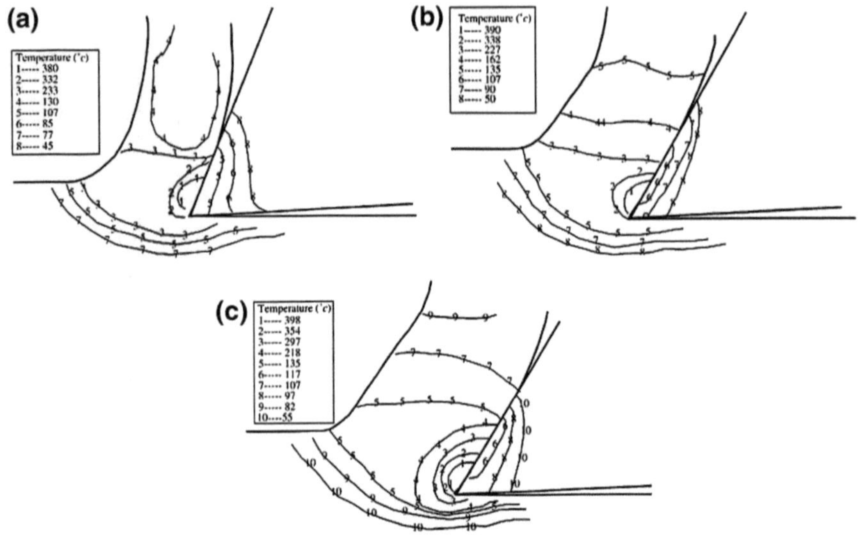

Fig. 1.20 Temperature distribution on the matrix. **a** Temperature contour for feed = 0.1 mm/rev. **b** Temperature contour for feed = 0.2 mm/rev. **c** Temperature contour for feed = 0.3 mm/rev (*source*: Zhu and Kishawy [104], with permission from Elsevier)

chip interface is substantially higher than that in the primary shear zone. At a cutting speed of 85 m/min and at feeds of 0.1, 0.2 and 0.3 mm/rev, the maximum temperatures predicted along the tool–chip interface were 380, 390 and 398°, respectively. According to the authors, the large amount of heat generated in the primary shear zone and the increase in the temperature are entirely related to the plastic deformation. An expected result is the higher temperatures with an increase in the feed.

Other studies on temperature predictions in machining of MMCs are concerned with the temperature observed on the cutting tool [43, 107]. The authors compared experimentally measured cutting temperatures to simulation predictions. The commercially available FEM solver ANSYS was used in the heat transfer analysis. Experiments were conducted on an A359 aluminum alloy reinforced with 20% by volume fraction of silicon carbide particles using a CVD-coated diamond insert. Cutting temperatures were measured using K-Type thermocouples along the locations schematically shown in Fig. 1.21.

A heat transfer model of the 3 D tool was constructed in ANSYS. In this model, transient heat conduction with adiabatic boundary conditions was applied for an initial machining temperature of 21°C. Convection boundary conditions with relative air flow caused by workpiece rotation were neglected as it showed negligible effects. At the tool–chip interface, the thermal boundary condition is uniformly specified as a heat flux, $\beta_r q_r$, where q_r is the overall heat flux at the rake face and β_r the heat partition index. The overall heat flux is due to the friction at the tool–chip interface and is determined by $(F_f/V_c)/A_c$, where F_f is the frictional

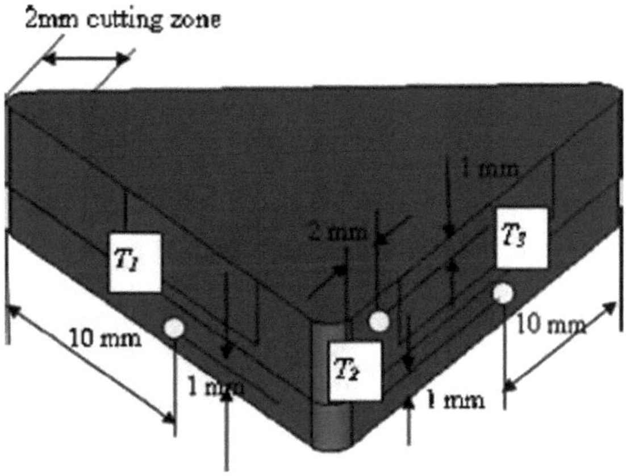

Fig. 1.21 Temperature measurement locations are marked as T_1, T_2 and T_3 of the K-Type thermocouples (*source*: Liu and Chou [43], with permission from Elsevier)

force, V_c is the chip sliding velocity and A_c is the tool–chip contact area, which are experimentally measured or determined through analytical models. The heat partition coefficient was numerically calculated by modeling a 2D chip model consisting of a long strip with chip thickness as the width dimension and the other dimension being relatively long to simulate an infinite domain. The model has a heat source, heat flux, and the tool–chip contact length, moving with the chip velocity along the boundary of the chip zone. The initial temperature of this chip model is the shear plane temperature calculated from theoretical models. In this manner the heat partition coefficient is determined numerically. Figure 1.22 shows the 3 D tool model used for predicting the tool temperatures. It was found that the cutting speed dominates the temperature rise of the cutting tool and the results qualitatively agreed with the temperature measurements from machining. The comparison of the simulated predictions with experimental measurements is given in Table 1.6.

1.6.1.4 3D Modeling of Sub-Surface Damage

The FEM machining model described in the previous section provides an insight into stress distributions when machining of an MMC. Practical machining is not a 2D process; therefore to circumvent this problem Dandekar and Shin [80] proposed a multi-step 3D FEM simulation for prediction of cutting forces and machining induced sub-surface damage in an A359/SiC/20p composite. The multi-step approach combines two modeling strategies: (a) a micromechanics-based approach and (b) an equivalent homogeneous material (EHM)-based approach. In the first step an EHM model is used for the overall prediction of cutting forces, temperature and the stress distributions in the composite undergoing machining (Fig. 1.23a). Three regions are chosen to calculate the sub-surface damage as marked on Fig. 1.23a. The actual locations of these regions are on the machined surface below the cutting tool near the marked areas. The resultant stress and

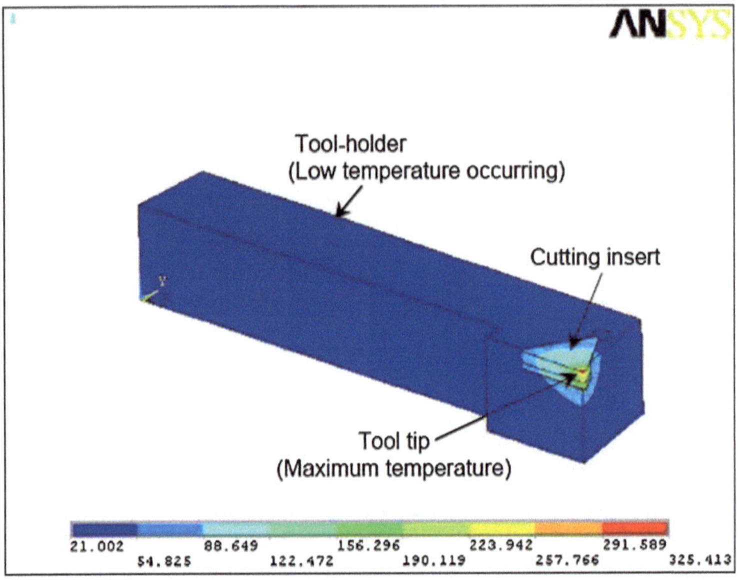

Fig. 1.22 Simulated cutting temperature distributions in the cutting tool (*source*: Chou and Liu [107], with permission from Elsevier)

Table 1.6 Comparisons between simulation and measured tool temperatures

		Simulation predictions (°C)			Experimental measurements (°C)		
Speed (m/min)	Feed (mm/rev)	T_1	T_2	T_3	T_1	T_2	T_3
60	0.1	84.1	72.5	72.9	66	65.5	71
60	0.3	99.9	87	87.2	77	90	97
180	0.1	102.4	89	89	87	91	99.5
180	0.3	97.1	81	81.3	95	81.5	87.5

Source: Liu and Chou [43], with permission from Elsevier

temperature from the 3D EHM model are subsequently imported into the local model (Fig. 1.23b. In the EHM model regions of 100 μm × 100 μm × 100 μm are selected along the cutting path at different locations in the workpiece. The post-processing is done in Third Wave Systems AdvantEdge^{TM} to obtain the stress and temperature distributions. The 3D microstructure of the local model is generated using the RSA (Random Sequential Adsorption) algorithm proposed by Rintoul and Torquato [108]. In the local model, the particle and matrix are assigned individual material properties, while the interface layer is modeled using 3D cohesive zone elements. Model details, boundary conditions, mesh size, convergence studies and validation of the material properties input in the FEM simulation can be obtained from Dandekar and Shin [80] and are not repeated here. In this manner the model harnesses the advantages of both the continuum (computational

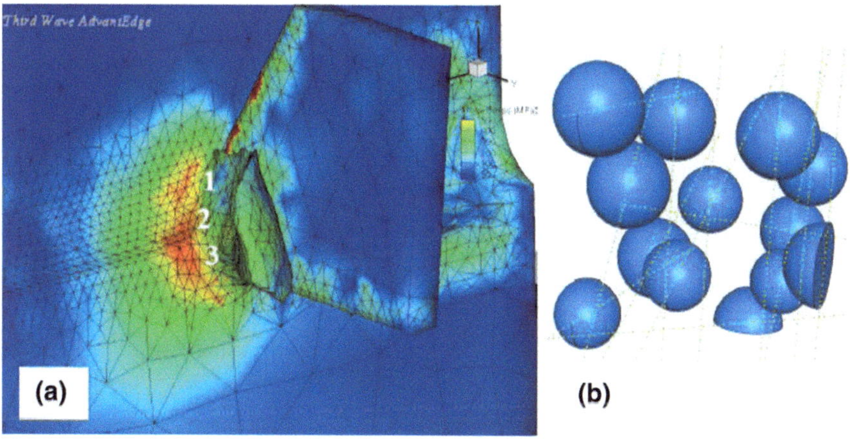

Fig. 1.23 **a** Stress distribution obtained from machining simulation software of a EHM MMC model using the 3D nose turning option in Third Wave Systems AdvantEdge code. **b** Multiparticle spheres in a random arrangement for 20% volume fraction (*source*: Dandekar and Shin [80], with permission from Elsevier)

speed and simplicity) and micromechanics (consideration of local effects) models, enabling it to accurately predict cutting forces and sub-surface damage.

The authors conducted machining tests on cast cylinders of A359/SiC/20p composites, supplied by MC-21 Inc. in the form of 68.5 mm diameter cylinders with a cut length of 152.4 mm. Each cutting test was repeated and the final plotted value of the cutting force Fig. 1.16 is an average of the steady state values for the two experiments conducted under the same cutting condition. For the measurement of damage, a scanning electron microscope was used to take five measurements from each of two different samples for all the cutting conditions. The reported value is the average value of these 10 measurements. A variation of 10–15% was observed in the experimental measurements.

At a cutting speed of 150 m/min and a feed rate of 0.05 mm/rev the simulated values of the cutting and thrust force were 78 and 56 N, respectively. On the other hand at a feed rate of 0.1 mm/rev the simulated cutting and thrust forces were found to be 126 and 55 N, respectively. The simulated cutting force and thrust force for the condition of 300 m/min cutting speed and a feed rate of 0.05 mm/rev were 80 and 56 N respectively. For the case of feed rate of 0.1 mm/rev the simulated results were 128 and 55 N, respectively. The coefficient of friction used in the simulations for all the cases was calculated from the experimental data.

On the whole the cutting forces as well as the thrust force match very well with experimental data. The trend observed in all the cases is similar such that the simulation under-predicts the cutting force by 7–8% and the thrust force by 6–12%. Figure 1.24 shows a representative comparison between the simulated and experimental data for a cutting speed of 300 m/min.

The measurement of post machining damage of MMC samples was done by obtaining SEM images. The images indicate the extent of debonding between the

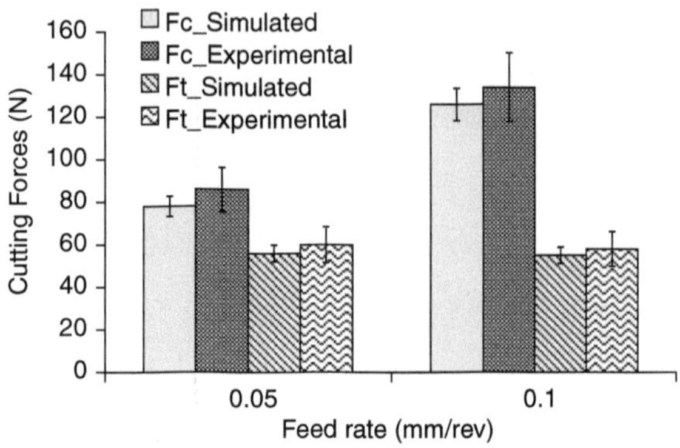

Fig. 1.24 Comparison of experimental and simulated results for cutting forces for machining at a cutting speed of 300 m/min (*source*: Dandekar and Shin [80], with permission from Elsevier)

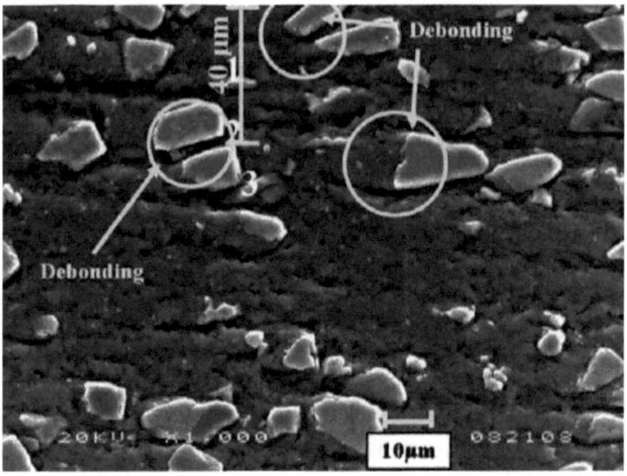

Fig. 1.25 Machined cross-section at cutting speed of 300 m/min and a feed rate of 0.05 mm/rev (*source*: Dandekar and Shin [80], with permission from Elsevier)

particles and the matrix along with particle fracture. The results indicate that the damage depth is primarily a function of feed rate. Figures 1.25 and 1.26 show the SEM images of the machined cross-section at the feed rate of 0.05 and 0.1 mm/rev at a cutting speed of 300 m/min, illustrating average sub-surface damage at 46 and 76 μm, respectively. As observed from the images it is clear that at the feed rate of 0.1 mm/rev there is a presence of higher damage, which corresponds to the higher-cutting forces observed during machining at higher feed rates. Regions of particle fracture can also be identified while machining at 300 m/min as marked on Fig. 1.26. Similar results were obtained from the SEM images for a cutting speed of 150 m/min at feed rates of 0.05 and 0.1 mm/rev, where the average sub-surface damage depth was 36 and 68 μm, respectively. As with the case of machining at 300 m/min the maximum damage was observed for a feed rate of 0.1 mm/rev.

Representative sectioned images of the simulated 100 μm cubic local damage model are shown in Figs. 1.27 and 1.28 for feed rates of 0.05 and 0.1 mm/rev, respectively at a cutting speed of 300 m/min, where the simulated von Mises stress distribution in MPa is shown. The regions of debonding and particle fracture are shown as insets of the regions where debonding and particle fracture occurred in Figs. 1.27 and 1.28. Similar simulation results were obtained for the cutting speed of

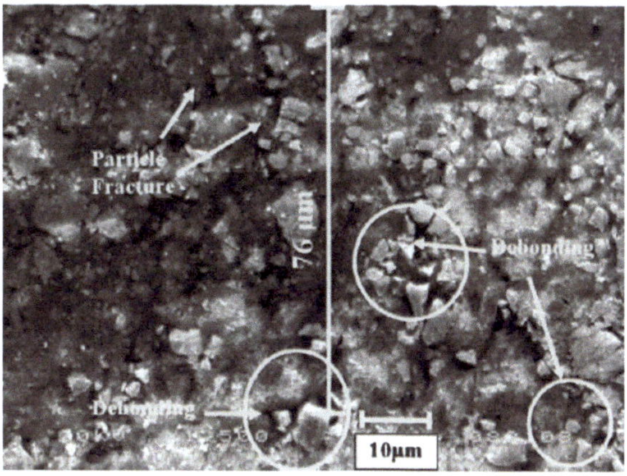

Fig. 1.26 Machined cross-section at cutting speed of 300 m/min and a feed rate of 0.1 mm/rev (*source*: Dandekar and Shin [80], with permission from Elsevier)

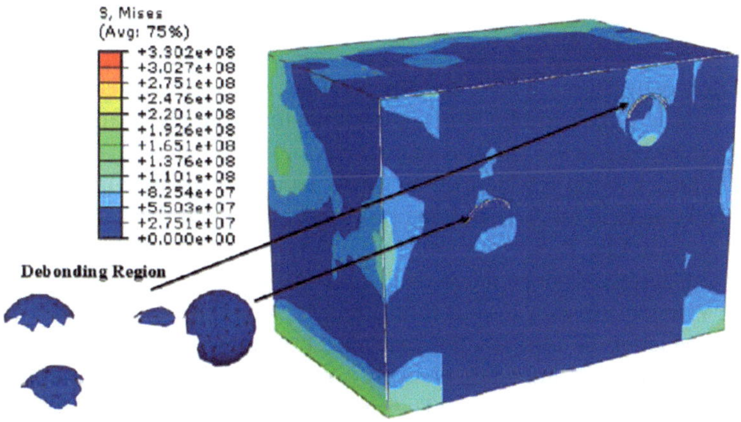

Fig. 1.27 Damage observed to a depth of 40.7 microns for a cutting speed of 300 m/min and a feed rate of 0.05 mm/rev (*source*: Dandekar and Shin [80], with permission from Elsevier)

150 m/min. Overall the experimental measurements compared very well with the simulated results, although the simulated values were slightly lower for all cases. This was consistent with the force comparisons where the thrust force was under predicted by the simulations. At a cutting speed of 150 m/min the simulated and experimental damage depths are 32 and 36 μm, respectively, for a feed rate of 0.05 mm/rev, while for a feed rate of 0.1 mm/rev the simulated and experimental measurements are 63 and 68 μm, respectively. At a cutting speed of 300 m/min and a feed rate of 0.05 and 0.1 mm/rev the simulated damage depths are 40.7 and 72 μm, respectively as shown in Figs. 1.27 and 1.28, while experimentally measured damage depths are 46 and 76 μm for a feed rate of 0.05 and 0.1 mm/rev, respectively.

During machining higher-cutting forces create more damage in terms of particle fracture and an increase in the debonding depth. This phenomenon is due to the

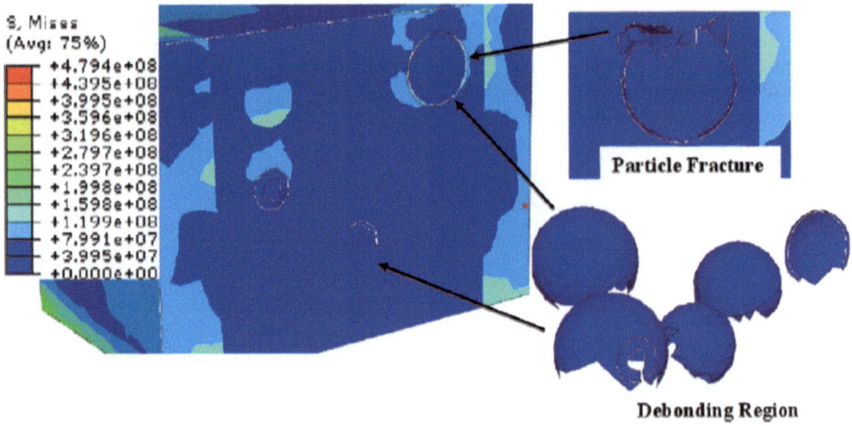

Fig. 1.28 Damage observed to a depth of 72 microns for a cutting speed of 300 m/min and a feed rate of 0.1 mm/rev (*source*: Dandekar and Shin [80], with permission from Elsevier)

interaction between the matrix and the particles. In the MMC material it is seen that the strain in the particles is much less than that observed in the matrix due to the high difference in the modulus of elasticity of the two phases. At higher-cutting forces there is a further increase in stress along the particle-matrix interface, resulting in particle fracture and deformation occurring at the bottom of the particle as seen in Fig. 1.28.

The sub-surface damage model includes capabilities for predicting debonding, particle fracture and matrix void formation. An advantage of the model is a significant reduction of computation time due to the multi-step simulation of the machining problem. A limitation of the model however, is the use of the EHM model in predicting the cutting forces. The treatment of the composite material as a homogenous material in the 3D model ignores the interaction of the tool with the particles, therefore ignoring the fluctuations in the cutting forces observed otherwise. The interaction of the particles with the tool has been studied by Pramanik et al. [19] for 2D orthogonal modeling. The effect of the particles is implicitly included in the homogenous material. To include this interaction in 3D modeling would be redundant and, hence for simplicity in applying the 3D machining model, can be neglected without losing accuracy. Nevertheless the treatment of machining simulation as a 3D nose turning results in incorporation of the effect of the tool nose radius and damage due to machining by the primary-and secondary-cutting edges.

1.6.2 Modeling Machining of Fiber Reinforced MMC

The machining simulation models presented above are all focused on particulate reinforced MMCs and there is a lack of models describing long-fiber MMC. Most of the models developed for machining of long-fiber composites deal with fiber

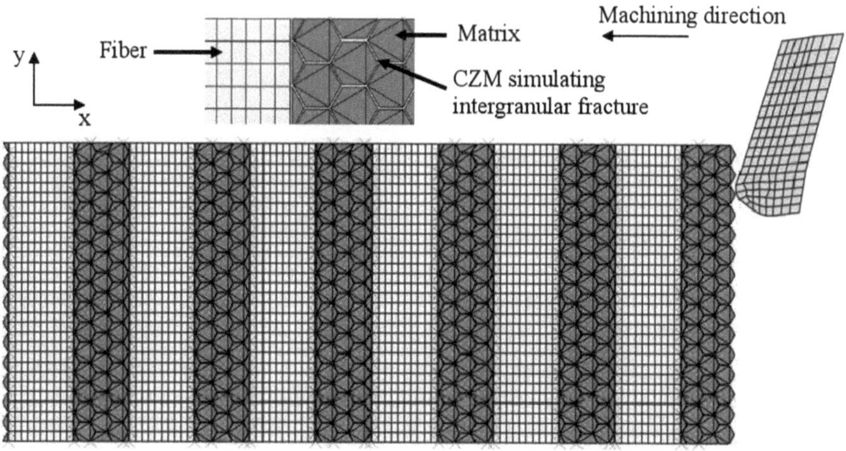

Fig. 1.29 Finite-element mesh for alumina fiber reinforced aluminum MMC (*source*: Dandekar and Shin [36], with permission from ASME)

reinforced polymer composites [72–74, 91]. All these authors indicated that the extent of damage depth and the cutting force increase with increasing fiber orientation. Utilizing a similar strategy to modeling machining of polymer composites, Dandekar and Shin [36] successfully simulated the machining of a long-fiber metal matrix composite.

In their model, a multi-phase FEM model has been implemented to study the effect of rake angle, material removal temperature and cutting speed on the machining of a long-fiber MMC [36]. The multi-phase approach utilizes a three phase finite-element mesh. The mesh is based on distinct properties of the fiber, matrix and the fiber-matrix interface and is shown in Fig. 1.29. The fiber is modeled as an anisotropic brittle material. The matrix on the other hand is considered as an elastic-plastic material modeled by a J–C model with damage in the matrix accounted for by a J–C damage model. The fiber-matrix interface and the grain boundaries are modeled using cohesive zone elements.

Simulations show that the primary method of failure in the matrix region is attributed to intergranular sliding with crack propagation achieved through failure of the cohesive zone elements. In the fiber, crack initiation and progression is accounted through the implementation of a damage law with the failure of the element resulting in the subsequent crack propagation. Figure 1.30 shows the progression of chip formation while machining of the MMC. The simulation allows the user to identify damage initiation sites, initiation of debonding at the fiber-matrix interface, fiber failure and subsequent fiber pullout. Damage in the composite after machining is characterized through excessive fiber breakage below the cutting plane. This is observed due to the brittle nature of the alumina fibers. To inspect damage after machining, SEM images were obtained by studying the interior cross-section of the workpieces. For each cutting condition, five measurements each from two different workpieces were

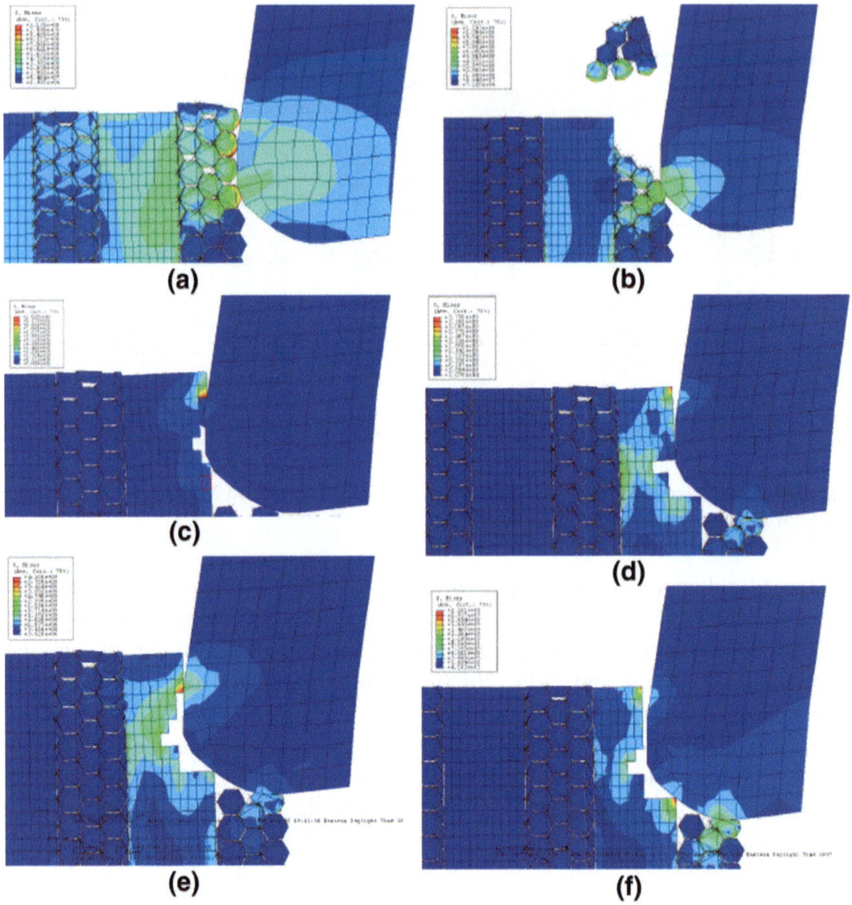

Fig. 1.30 Simulated results $Vc = 30$ m/min, $f = 0.02$ mm/rev, a tool rake angle of 5° machining of a long-fiber MMC

performed. The reported value is the average value of these 10 measurements. A variation of 20% was observed in the experimental measurements. The SEM image indicates the extent of debonding between the fiber and the matrix, microcracking of fibers and fiber pullout. The indicator for fiber pullout is the presence of matrix-rich regions after conventional machining of workpieces. The damage in the composite decreased with decreasing thrust force. A representative image for machining at a cutting speed of 30 m/min, feed of 0.02 mm/rev and depth of cut of 0.5 is shown in Fig. 1.31. A simulation result showing the fiber damage and debonding is presented in Fig. 1.32. The average measured value for the debonding/fiber damage is 157 ± 15 μm, while simulations predict a damage depth of 162 ± 2 μm. Further validations of the presented model can be obtained from Dandekar and Shin [36].

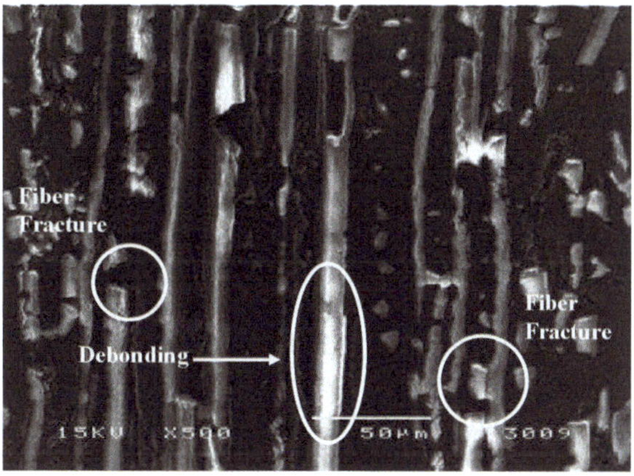

Fig. 1.31 Sub-surface damage measurement at $f = 0.02$ mm/rev $V_C = 30$ m/min, $d = 0.5$ mm for conventional machining (*source*: Dandekar and Shin [36], with permission from ASME)

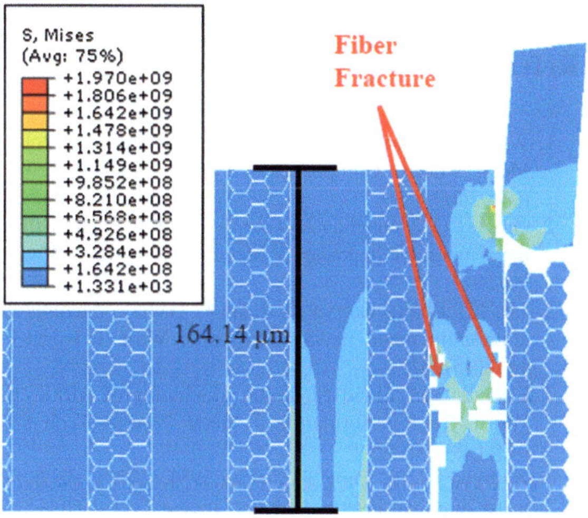

Fig. 1.32 Simulated results $Vc = 30$ m/min, $f = 0.02$ mm/rev, a tool rake angle of 5° for conventional machining (*source*: Dandekar and Shin [36], with permission from ASME)

1.7 Conclusion

In this chapter, the mechanics of chip formation in machining (turning) of particulate and fiber reinforced MMCs was introduced. Analytical models for determining the cutting forces and cutting temperatures for machining of particulate reinforced MMC's were introduced. The cutting forces, surface quality and tool wear are mainly controlled by the reinforcement size, feed and cutting speed. Poly crystalline diamond (PCD) is the most suitable tool material for machining of MMCs. The most common chip type produced during machining of particulate MMC's is the sawtooth type of chip. On the other hand machining of fiber reinforced composites produced chips which are extremely short and discontinuous. Most of the studies are involved with machining of particulate reinforced MMCs, while machining studies on long-fiber reinforced MMCs are highly lacking. Analytical models were presented to understand the cutting mechanism for both types of composites. The analytical models are capable of predicting the cutting forces with good accuracy, albeit they fail to account for all the various deformation phenomena observed in machining of MMCs. Analytical predictions of the cutting temperatures in machining of MMCs deviate from experimental measurements. To this end 2D and 3D finite element models were introduced to explain the chip formation process, tool–particle interaction, prediction of cutting forces, cutting temperatures and the sub-surface damage. Good agreements have been found between model predictions and experimental measurements of the cutting forces, cutting temperatures tool–chip interaction and sub-surface damage for machining of MMCs. The analytical and numerical models presented in this chapter assist in selecting machining parameters: tool geometry and cutting conditions to improve the machinability of MMC.

References

1. Weinert K, König W (1993) A consideration of tool wear mechanism when machining Metal Matrix Composites (MMC). Ann CIRP 42(1):95–98. doi:10.1016/s0007-8506(07)62400-7
2. Quigley O, Monaghan J, O'Reilly P (1994) Factors affecting the machinability of an Al/SiC metal-matrix composite. J Mater Process Tech 43(1):21–36. doi:10.1016/0924-0136(94)90159-7
3. Lin JT, Bhattacharyya D, Lane C (1995) Machinability of a silicon carbide reinforced aluminium metal matrix composite. Wear 181–183(Part 2):883–888. doi:10.1016/0043-1648(95)90211-2
4. El-Gallab M, Sklad M (1998) Machining of Al/SiC particulate metal-matrix composites. Part I: Tool performance. J Mater Process Tech 83(1–3):151–158. doi:10.1016/s0924-0136(98)00054-5
5. Andrewes CJE, Feng H-Y, Lau WM (2000) Machining of an aluminum/SiC composite using diamond inserts. J Mater Process Tech 102(1–3):25–29. doi:10.1016/s0924-0136(00)00425-8
6. El-Gallab M, Sklad M (2000) Machining of Al/SiC particulate metal matrix composites. Part III: Comprehensive tool wear models. J Mater Process Tech 101(1–3):10–20. doi:10.1016/s0924-0136(99)00351-9
7. Quan Y, Zhou Z (2000) Tool wear and its mechanism for cutting SiC particle-reinforced aluminium matrix composites. J Mater Process Tech 100(1–3):194–199. doi:10.1016/s0924-0136(99)00405-7
8. El-Gallab M, Sklad M (1998) Machining of Al/SiC particulate metal matrix composites. Part II: Workpiece surface integrity. J Mater Process Tech 83(1–3):277–285. doi:10.1016/s0924-0136(98)00072-7
9. Manna A, Bhattacharayya B (2003) A study on machinability of Al/SiC-MMC. J Mater Process Tech 140(1–3):711–716. doi:10.1016/s0924-0136(03)00905-1
10. Pramanik A, Zhang LC, Arsecularatne JA (2008) Machining of metal matrix composites: effect of ceramic particles on residual stress, surface roughness and chip formation. Int J Mach Tool Manu 48(15):1613–1625. doi:10.1016/j.ijmachtools.2008.07.008
11. Manna A, Bhattacharyya B (2004) Investigation for optimal parametric combination for achieving better surface finish during turning of Al/SiC-MMC. Int J Adv Manuf Tech 23:658–665. doi:10.1007/s00170-003-1624-z

12. Ozben T, Kilickap E, ÇakIr O (2008) Investigation of mechanical and machinability properties of SiC particle reinforced Al-MMC. J Mater Process Tech 198(1–3):220–225. doi:10.1016/j.jmatprotec.2007.06.082
13. Wang J, Huang CZ, Song WG (2003) The effect of tool flank wear on the orthogonal cutting process and its practical implications. J Mater Process Tech 142(2):338–346. doi: 10.1016/s0924-0136(03)00604-6
14. Manna A, Bhattacharayya B (2005) Influence of machining parameters on the machinability of particulate reinforced Al/SiC-MMC. Int J Adv Manuf Tech 25:850–856. doi:10.1007/s00170-003-1917-2
15. Kannan S, Kishawy HA, Balazinski M (2006) Flank wear progression during machining metal matrix composites. J Manuf Sci E-T ASME 128:787–791. doi:10.1115/1.2164508
16. Muthukrishnan N, Murugan M, Prahlada Rao K (2008) Machinability issues in turning of Al-SiC (10 p) metal matrix composites. Int J Adv Manuf Tech 39:211–218. doi:10.1007/s00170-007-1220-8
17. Davim JP (2002) Diamond tool performance in machining metal-matrix composites. J Mater Process Tech 128(1–3):100–105. doi:10.1016/s0924-0136(02)00431-4
18. Joshi SS, Ramakrishnan N, Nagarwalla HE, Ramakrishnan P (1999) Wear of rotary carbide tools in machining of Al/SiCp composites. Wear 230(2):124–132. doi:10.1016/s0043-1648(99)00092-7
19. Pramanik A, Zhang LC, Arsecularatne JA (2007) An FEM investigation into the behaviour of metal matrix composites: tool-particle interaction during orthogonal cutting. Int J Mach Tool Manu 47(10):1497–1506. doi:10.1016/j.ijmachtools.2006.12.004
20. Kishawy HA, Kannan S, Balazinski M (2004) An energy based analytical force model for orthogonal cutting of metal matrix composites. Ann CIRP 53(1):91–94. doi:10.1016/s0007-8506(07)60652-0
21. Davim JP (2003) Design of optimisation of cutting parameters for turning metal matrix composites based on the orthogonal arrays. J Mater Process Tech 132(1–3):340–344. doi: 10.1016/s0924-0136(02)00946-9
22. Monaghan J, Brazil D (1997) Modeling the sub-surface damage associated with the machining of a particle reinforced MMC. Comp Mater Sci 9(1–2):99–107. doi:10.1016/s0927-0256(97)00063-3
23. El-Gallab MS, Sklad MP (2004) Machining of aluminum/silicon carbide particulate metal matrix composites. Part IV: Residual stresses in the machined workpiece. J Mater Process Tech 152(1):23–34. doi:10.1016/j.jmatprotec.2004.01.061
24. Davim JP, Baptista AM (2000) Relationship between cutting force and PCD cutting tool wear in machining silicon carbide reinforced aluminum. J Mater Process Tech 103(3): 417–423. doi:10.1016/s0924-0136(00)00495-7
25. Chambers AR (1996) The machinability of light alloy MMCs. Composites Part A-Appl S 27(2):143–147. doi:10.1016/1359-835x(95)00001-i
26. Ciftci I, Turker M, Seker U (2004) Evaluation of tool wear when machining SiCp-reinforced Al-2014 alloy matrix composites. Mater Des 25(3):251–255. doi:10.1016/j.matdes.2003.09.019
27. Ding X, Liew WYH, Liu XD (2005) Evaluation of machining performance of MMC with PCBN and PCD tools. Wear 259(7–12):1225–1234. doi:10.1016/j.wear.2005.02.094
28. Tomac N, Tannessen K, Rasch FO (1992) Machinability of particulate aluminium matrix composites. Ann CIRP 41(1):55–58. doi:10.1016/s0007-8506(07)61151-2
29. Arsecularatne JA, Zhang LC, Montross C (2006) Wear and tool life of tungsten carbide, PCBN and PCD cutting tools. Int J Mach Tool Manu 46(5):482–491. doi:10.1016/j.ijmachtools.2005.07.015
30. Hung NP, Zhong CH (1996) Cumulative tool wear in machining metal matrix composites. Part I: Modelling. J Mater Process Tech 58(1):109–113. doi:10.1016/0924-0136(95)02114-0
31. Heath PJ (2001) Developments in applications of PCD tooling. J Mater Process Tech 116(1):31–38. doi:10.1016/s0924-0136(01)00837-8

32. Hyde AR (1990) Ceramic matrix composites. Mater Des 11(1):30–36. doi:10.1016/0261-3069(90)90087-z
33. Komanduri R (1997) Machining of fiber-reinforced composites. Mach Sci Technol 1:113–152
34. Varadarajan YS, Vijayaraghavan L, Krishnamurthy R (2002) The machinability characteristics of aluminosilicate fiber reinforced Al alloy composite. Mater Manuf Process 17:811–824. doi:10.1081/AMP-120016059
35. Weinert K, Lange M (2003) Machining of fibre reinforced magnesium. In: Proceedings of the second Osaka international conference on platform science and technology for advanced magnesium alloys 2003, Jan 26, 2003–Jan 30, 2003, Osaka, Japan. Materials science forum. Trans Tech Publications Ltd, pp 823–828
36. Dandekar CR, Shin YC (2010) Laser-assisted machining of a fiber reinforced metal matrix composite. J Manuf Sci E-T ASME 132(6):061004–061008. doi:10.1115/1.4002548
37. Davim JP (2007) Application of merchant theory in machining particulate metal matrix composites. Mater Des 28(10):2684–2687. doi:10.1016/j.matdes.2006.10.015
38. Merchant ME (1945) Mechanics of the metal cutting process. I. Orthogonal cutting and a type 2 chip. J Appl Phys 16:267–275
39. Waldorf DJ (2006) A simplified model for ploughing forces in turning. J Manuf Processes 8:76–82
40. Yan C, Zhang L (1994) Single-point scratching of 6061 Al alloy reinforced by different ceramic particles. Appl Compos Mater 1:431–447
41. Zhang LC (2009) Cutting composites: a discussion on mechanics modelling. J Mater Process Tech 209(9):4548–4552. doi:10.1016/j.jmatprotec.2008.10.023
42. Boothroyd G, Knight WA (2005) Fundamentals of machining and machine tools. crc mechanical engineering, 3rd edn. CRC Pr I Llc, Boca Raton, Florida
43. Liu J, Chou YK (2007) Cutting tool temperature analysis in heat-pipe assisted composite machining. J Manuf Sci E-T ASME 129:902–910. doi:10.1115/1.2752528
44. Weiner JH (1955) Shear-plane temperature distribution in orthogonal cutting. T ASME 77(8):1331–1336
45. Rapier AC (1954) A theoretical investigation of the temperature distribution in the metal cutting process. Brit J Appl Phys 5(11):400–405. doi:10.1088/0508-3443/5/11/306
46. Kalpakjian S (1996) Manufacturing processes for engineering materials. Addison-Wesley, Reading, MA
47. Karthikeyan R, Ganesan G, Nagarazan RS, Pai BC (2001) A critical study on machining of Al/SiC composites. Mater Manuf Process 16:47–60. doi:10.1081/AMP-100103696
48. Monaghan J, Brazil D (1998) Modelling the flow processes of a particle reinforced metal matrix composite during machining. Composites Part A-Appl S 29(1–2):87–99. doi:10.1016/s1359-835x(97)00047-x
49. Hung NP, Yeo SH, Lee KK, Ng KJ (1998) Chip formation in machining particle-reinforced metal matrix composites. Mater Manuf Process 13:85–100
50. Hung NP, Loh NL, Venkatesh VC (1999) Machining of metal matrix composites. In: Jahanmir S, Ramulu M, Koshy P (eds) Machining of ceramics and composites. Marcel Dekker Inc, New York, pp 295–356
51. Joshi SS, Ramakrishnan N, Ramakrishnan P (2001) Micro-structural analysis of chip formation during orthogonal machining of Al/SiCp composites. J Eng Mater-T ASME 123:315–321. doi:10.1115/1.1356026
52. Koplev A, Lystrup A, Vorm T (1983) The cutting process, chips, and cutting forces in machining CFRP. Composites 14(4):371–376. doi:10.1016/0010-4361(83)90157-x
53. Takeyama H, Iijima N (1988) Machinability of glassfiber reinforced plastics and application of ultrasonic machining. Ann CIRP 37(1):93–96. doi:10.1016/s0007-8506(07)61593-5
54. Kim KS, Lee DG, Kwak YK, Namgung S (1992) Machinability of carbon fiber-epoxy composite materials in turning. J Mater Process Tech 32(3):553–570. doi:10.1016/0924-0136(92)90253-o

55. Nayak D, Bhatnagar N, Mahajan P (2005) Machining studies of uni-directional glass fiber reinforced plastic (UD-GFRP) composites. Part 1: Effect of geometrical and process parameters. Mach Sci Technol 9:481–501. doi:10.1080/10910340500398167
56. Wang XM, Zhang LC (1999) Machining damage in unidirectional fiber-reinforced plastics. In: Wan J, Scott W, Zhang LC (eds) Abrasive technology—current development and applications. World Scientific, Singapore, pp 429–436
57. Zhang LC, Zhang HJ, Wang XM (2001) A force prediction model for cutting unidirectional fibre-reinforced plastics. Mach Sci Technol 5:293–305. doi:10.1081/MST-100108616
58. Bergman F, Jacobson S (1994) Tool wear mechanisms in intermittent cutting of metal matrix composites. Wear 179(1–2):89–93. doi:10.1016/0043-1648(94)90224-0
59. Mackerle J (1998) Finite-element analysis and simulation of machining: a bibliography (1976–1996). J Mater Process Tech 86(1–3):17–44. doi:10.1016/s0924-0136(98)00227-1
60. Mackerle J (2003) Finite element analysis and simulation of machining: an addendum: a bibliography (1996–2002). Int J Mach Tool Manu 43(1):103–114. doi:10.1016/s0890-6955(02)00162-1
61. Soo SL, Aspinwall DK (2007) Developments in modeling of metal cutting processes. In: IMechE, vol. 221, Part L: J Mater Design Appl, pp 197–211
62. ABAQUS User's Manual, Version 6.5 (2004) Hibbitt, Karlsson & Sorensen, Inc., Pawtucket, RI
63. Olovsson L, Nilsson L, Simonsson K (1999) An ALE formulation for the solution of two-dimensional metal cutting problems. Comput Struct 72(4–5):497–507. doi:10.1016/s0045-7949(98)00332-0
64. Movahhedy MR, Altintas Y, Gadala MS (2002) Numerical analysis of metal cutting with chamfered and blunt tools. J Manuf Sci E-T ASME 124:178–188. doi:10.1115/1.1445147
65. Movahhedy M, Gadala MS, Altintas Y (2000) Simulation of the orthogonal metal cutting process using an arbitrary Lagrangian-Eulerian finite-element method. J Mater Process Tech 103(2):267–275. doi:10.1016/s0924-0136(00)00480-5
66. Tian Y, Shin YC (2004) Finite element modeling of machining of 1020 steel including the effects of round cutting edge. In: Papers Presented at NAMRC 32, June 1, 2004–June 4, 2004, Charlotte, NC, United states, Transactions of the North American Manufacturing Research Institute of SME. Society of Manufacturing Engineers, pp 111–118
67. Özel T, Zeren E (2005) Finite element method simulation of machining of AISI 1045 Steel with a Round Edge Cutting Tool. In: Proceedings of 8th CIRP international workshop on modeling of machining operations. Chemnitz, Germany
68. Ozel T, Zeren E (2007) Finite element modeling the influence of edge roundness on the stress and temperature fields induced by high-speed machining. Int J Adv Manuf Tech 35:255–267. doi:10.1007/s00170-006-0720-2
69. Haglund AJ, Kishawy HA, Rogers RJ (2008) An exploration of friction models for the chip-tool interface using an Arbitrary Lagrangian-Eulerian finite element model. Wear 265(3–4):452–460. doi:10.1016/j.wear.2007.11.025
70. Dandekar CR, Shin YC, Barnes J (2010) Machinability improvement of titanium alloy (Ti-6Al-4 V) via LAM and hybrid machining. Int J Mach Tools Manuf 50(2):174–182. doi:10.1016/j.ijmachtools.2009.10.013
71. Camus G (2000) Modelling of the mechanical behavior and damage processes of fibrous ceramic matrix composites: application to a 2D SiC/SiC. Int J Solids Struct 37(6):919–942. doi:10.1016/s0020-7683(99)00065-7
72. Arola D, Ramulu M (1997) Orthogonal cutting of fiber-reinforced composites: a finite element analysis. Int J Mech Sci 39(5):597–613. doi:10.1016/s0020-7403(96)00061-6
73. Arola D, Sultan MB, Ramulu M (2002) Finite element modeling of edge trimming fiber reinforced plastics. J Manuf Sci E-T ASME 124:32–41
74. Nayak D, Bhatnagar N, Mahajan P (2005) Machining studies of ud-frp composites. Part 2: Finite element analysis. Mach Sci Technol 9:503–528. doi:10.1080/10910340500398183

75. Rao GVG, Mahajan P, Bhatnagar N (2007) Machining of UD-GFRP composites chip formation mechanism. Compos Sci Technol 67(11–12):2271–2281. doi:10.1016/j.compscitech.2007.01.025
76. Rao GVG, Mahajan P, Bhatnagar N (2007) Micro-mechanical modeling of machining of FRP composites—cutting force analysis. Compos Sci Technol 67(3–4):579–593. doi:10.1016/j.compscitech.2006.08.010
77. Johnson GR, Cook WH (1983) A constitutive model and data for metals subjected to large strains, high rates and high temperatures. In: Proceedings of the seventh international symposium on ballistics. The Hague, the Netherlands, pp 541–547
78. Norton FH (1929) Creep of steel at high temperature. McGraw-Hill, New York
79. HOFF NJ (1954) Approximate analysis of structures in the presence of moderately large creep deformations. Q Appl Math 12:49–55
80. Dandekar CR, Shin YC (2009) Multi-step 3D finite element modeling of subsurface damage in machining particulate reinforced metal matrix composites. Composites Part A: Appl Sci Manuf 40(8):1231–1239. doi:10.1016/j.compositesa.2009.05.017
81. Zhou L, Huang ST, Wang D, Yu XL (2011) Finite element and experimental studies of the cutting process of SiCp/Al composites with PCD tools. Int J Adv Manuf Tech 52:619–626. doi:10.1007/s00170-010-2776-2
82. Macdougall DAS, Harding J (1999) A constitutive relation and failure criterion for Ti6Al4 V alloy at impact rates of strain. J Mech Phys Solids 47(5):1157–1185. doi:10.1016/s0022-5096(98)00086-6
83. Follansbee PS, Gray GT III (1989) An analysis of the low temperature, low and high strain-rate deformation of Ti-6Al-4 V. Metall Trans A 20A:863–874
84. Nemat-Nasser S, Guo W-G, Nesterenko VF, Indrakanti SS, Gu Y-B (2001) Dynamic response of conventional and hot isostatically pressed Ti-6Al-4 V alloys: experiments and modeling. Mech Mater 33:425–439. doi:10.1016/S0167-6636(01)00063-1
85. Lemaitre J, Desmorat R (2005) Engineering damage mechanics: ductile, creep, fatigue and brittle failures. Springer-Verlag, Berlin, Heidelberg
86. Marusich TD, Ortiz M (1995) Simulation of chip formation in high-speed machining. In: Proceedings of the 1995 joint ASME applied mechanics and materials conference, Machining of Advanced Materials, AMD, pp 127–139
87. Cantonwine PE (2003) Strength of thermally exposed alumina fibers. Part I: Single filament behavior. J Mater Sci 38:461–470. doi:10.1023/A:1021867530979
88. 3M (2004) Nextel™ Ceramic textilestechnical notebook. St. Paul, MN
89. Bansal NP (2005) Handbook of ceramic composites. Kluwer Academic Publishers, New York
90. Orifici AC, Herszberg I, Thomson RS (2008) Review of methodologies for composite material modelling incorporating failure. Comput Struct 86(1–3):194–210. doi:10.1016/j.compstruct.2008.03.007
91. Dandekar CR, Shin YC (2008) Multiphase finite element modeling of machining unidirectional composites: prediction of debonding and fiber damage. J Manuf Sci E-T ASME 130:0510161–05101612. doi:10.1115/1.2976146
92. Needleman A (1987) A continuum model for void nucleation by inclusion debonding. J Appl Mech-T ASME 54:525–531
93. Tvergaard V (1990) Effect of fibre debonding in a whisker-reinforced metal. Mater Sci Eng 125(2):203–213. doi:10.1016/0921-5093(90)90170-8
94. Xu XP, Needleman A (1995) Numerical simulations of dynamic crack growth along an interface. Int J Fracture 74:289–324
95. Camacho GT, Ortiz M (1996) Computational modelling of impact damage in brittle materials. Int J Solids Struct 33(20–22):2899–2938. doi:10.1016/0020-7683(95)00255-3
96. Chandra N, Li H, Shet C, Ghonem H (2002) Some issues in the application of cohesive zone models for metal-ceramic interfaces. Int J Solids Struct 39(10):2827–2855. doi:10.1016/s0020-7683(02)00149-x

97. Xie JQ, Bayoumi AE, Zbib HM (1998) FEA modeling and simulation of shear localized chip formation in metal cutting. Int J Mach Tool Manu 38(9):1067–1087. doi:10.1016/s0890-6955(97)00063-1
98. Bäker M, Rösler J, Siemers C (2002) A finite element model of high speed metal cutting with adiabatic shearing. Comput Struct 80(5–6):495–513. doi:10.1016/s0045-7949(02)00023-8
99. Simoneau A, Ng E, Elbestawi MA (2006) Chip formation during microscale cutting of a medium carbon steel. Int J Mach Tool Manu 46(5):467–481. doi:10.1016/j.ijmachtools.2005.07.019
100. Shet C, Deng X (2000) Finite element analysis of the orthogonal metal cutting process. J Mater Process Tech 105(1–2):95–109. doi:10.1016/s0924-0136(00)00595-1
101. Ceretti E, Lucchi M, Altan T (1999) FEM simulation of orthogonal cutting: serrated chip formation. J Mater Process Tech 95(1–3):17–26. doi:10.1016/s0924-0136(99)00261-7
102. Hua J, Shivpuri R (2002) Influence of crack mechanics on the chip segmentation in the machining of Ti-6Al-4 V. In: Proceedings of the 9th ISPE international conference on concurrent engineering, Cranfield, UK, pp 357–365
103. Filice L, Micari F, Rizzuti S, Umbrello D (2007) A critical analysis on the friction modelling in orthogonal machining. Int J Mach Tool Manu 47(3–4):709–714. doi:10.1016/j.ijmachtools.2006.05.007
104. Zhu Y, Kishawy HA (2005) Influence of alumina particles on the mechanics of machining metal matrix composites. Int J Mach Tool Manu 45(4–5):389–398. doi:10.1016/j.ijmachtools.2004.09.013
105. Quan Y, Ye B (2003) The effect of machining on the surface properties of SiC/Al composites. J Mater Process Tech 138(1–3):464–467. doi:10.1016/s0924-0136(03)00119-5
106. Reddy PR, Sriramakrishna AA (2002) Analysis of orthogonal cutting of aluminum-based composites. Defence Sci J 52(4):375–382
107. Chou YK, Liu J (2005) CVD diamond tool performance in metal matrix composite machining. Surf Coat Tech 200(5–6):1872–1878. doi:10.1016/j.surfcoat.2005.08.094
108. Rintoul MD, Torquato S (1997) Reconstruction of the structure of dispersions. J Colloid Interf Sci 186(2):467–476. doi:10.1006/jcis.1996.4675

Chapter 2
Surface Integrity When Machining Metal Matrix Composites

Abdul B. Sadat

Abstract Traditional machining operations normally lead to the alterations of surface and subsurface that makes it different from the bulk material. In order to avoid undesirable alterations that may have adverse affect on the quality of the machined components it is essential to know the various types of alteration and their origin. It is also important to know the cutting parameters and tool geometry that will lead to machined components of high-quality.

2.1 Introduction

In spite of the rapid advances being made in the development of new and improved production techniques, such as precision casting, directional solidification, and net shape forming, traditional metal machining processes still feature directly or indirectly in the manufacture of most of the items in our present technological society. It is generally accepted that machining introduces changes into the surface region that makes it different from the bulk materials. As mechanical properties such as fatigue, creep, and stress corrosion cracking are important material properties that are highly surface sensitive, it is important to obtain complete information on the surface characteristics of machined components, in addition to those mechanical and physical properties generally considered essential. It is also important to identify the cutting parameters that will generate a high-quality surface at minimum cost. It has long been evident that the traditional machining

A. B. Sadat (✉)
Industrial and Manufacturing Department, California State Polytechnic University, Pomona, CA 91768, USA
e-mail: absadat@csupomona.edu

J. P. Davim (ed.), *Machining of Metal Matrix Composites*,
DOI: 10.1007/978-0-85729-938-3_2,

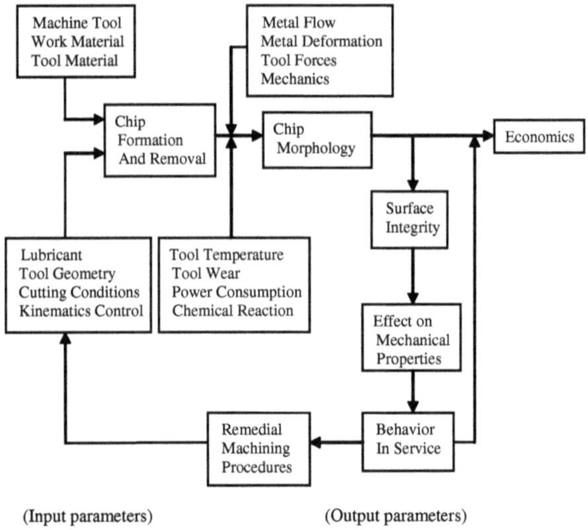

Fig. 2.1 A simplified machining system

process, although conceptually simple, is perhaps one of the most complex of the manufacturing operations employed today. A complete description of the process involves the consideration of an entire system, consisting of various input and output parameters linked together through a variety of complex internal phenomena. A simple scheme showing the relationship between some of the more important parameters involved in machining process is given in Fig. 2.1. A change in any one of the input parameters can lead to a change in all the output parameters.

2.2 Surface Integrity

Researchers in the field of machining have generally accepted the term surface integrity to describe the nature or condition of the machined surface region. It is interpreted as those elements that describe the actual structure of both surface and subsurface [1]. If the surface of a machined component posses integrity, then it is not affected by the action of the machining process used in its fabrication. Conversely, if integrity is lacking, then surface and subsurface damage have occurred as a consequence of the impact of machining process. Surface integrity has two distinct and important aspects, (1) surface topography that describes surface roughness and other features associated with the geometry of the surface, and (2) surface region metallurgy of the layer that is produced by the machining process including the effect of any alterations with respect to base material.

A list of more obvious elements that can be used to describe surface integrity is given in Table 2.1. Clearly, this method of division is by no means unique and other methods could be devised. In addition, not all elements mentioned here can

Table 2.1 Elements of surface integrity

Surface elements	Subsurface elements
Surface roughness, Waviness, Texture, Distortion, Micro-cracks, Macro-cracks, Tears, Laps, Pits, Cavities, Surface corrosion, Intergranular corrosion, Debris, Hardness variations, Surface phase transformation, Surface structural changes, Thermodynamic property changes, Grain boundaries, Stress induced surface roughness, Variation in electrostatic potential, Residual stress	Micro-cracks, Macro-cracks, Phase transformation, Compositional changes, Hardness changes, Plastic deformation, Dislocation density and distribution, Recovery, Recrystallization, Grain growth, Residual stresses, Inclusions, Voids, Vacancies interstitials, Frenkel defects, Solute atoms, Twins, Stacking fault, Antiphase boundaries, Domain walls

be expected in a given material system. Some phenomena referred to are not possible in many materials. For example, domain walls would be expected only in ferro-magnetic or ferro-electric materials. Also, surface and subsurface alterations need not necessarily affect adversely the aforementioned surface sensitive mechanical properties. Indeed some forms of alteration can prove beneficial. For example, the introduction of residual compressive stresses in the surface region of a component can lead to substantial improvement in fatigue life. It is for this reason that critical surfaces are often given post-machining treatments, such as shot peening in order to improve fatigue life.

2.3 Surface Integrity Evaluation Techniques

An excellent early review of techniques for assessing surface integrity is presented in Ref. [2]. Table 2.2 lists a summary of various methods used in identifying various defects produced as a result of machining. In this section a few of the most common techniques used in evaluating the quality of the machined surface region of metal matrix composites namely, optical and scanning electron microscopy of surface and subsurface, plastic deformation of surface and subsurface, residual stress analysis of the surface region, and surface topography will be presented.

2.3.1 Optical and Scanning Electron Microscopy

Optical and scanning electron microscopies of machined surface are commonly used to assess the quality of the machined surface. Machined surfaces are examined using optical microscope at various magnification to investigate the presence of surface damage in the form of cavities, macrocracks, scratch marks, etc.

Table 2.2 Techniques for assessing surface integrity

Interest area	Common techniques	Identifiable defects
Surface topography	Profilometry, Optical microscopy, Interference microscopy, Electron microscopy, Metrology	Surface profile, Surface roughness, Waviness, Lay, Surface defects, Distortion
Surface structure	Visual inspection, Optical microscopy, Dye penetrant, Magnetic particle, Microetch eddy current, Ultrasonic pulse echo, Ultrasonic velocity, High frequency ultrasonic, X-ray spectrography, Radioactive gas penetrant, Surface electrical resistance, Beta backscatter, Electrochemical potential, Microhardness, Deflection-etching	Surface defect, Crack detection, Surface phase transformation, Surface hardness changes, Plastic deformation, Surface residual stress, Surface compositional changes
Surface chemistry	Laser probe mass spectrometry, X-ray spectroscopy, Auger spectroscopy, X-ray microprobe analysis, Transmission electron microscopy	Surface composition
Subsurface structure	Optical microscopy, Eddy current, Ultrasonic pulse echo, X-ray radiography, Gamma radiography, Radioactive gas penetrant, X-ray diffraction, Deflection-etching, Microhardness testing, Scanning electron microscopy, Transmission electron microscopy	Subsurface defects, Phase transformations, Recrystallization, Grain growth, Hardness variations, Inclusions, Residual stresses, Dislocation structures
Subsurface	Chemical analysis, X-ray spectrometry, X-ray diffraction and X-ray microprobe analysis, Auger spectroscopy, Transmission electron microscopy	Subsurface composition

Fig. 2.2 An optical photomicrograph of machined surface (×100), hot-rolled Al/SiC, 20% vol. SiC

Selected samples are then used for detailed analysis and for observing the presence of types of damages that are not possible by using optical microscope such as formation of voids, microcracks, etc. Figure 2.2 is a typical optical photomicrograph of the machined surface showing long grooves parallel to the direction of cutting velocity and pitted area. The equally spaced long grooves are attributed to feed marks caused by the geometry of the cutting tool.

A typical scanning electron photomicrograph of the machined surface is presented in Fig. 2.3. Surface damage in the form of short and long grooves parallel to the direction of cutting, pitted area, and cavities are generally present and randomly occurs when machining metal matrix composites. It is understood that during cutting operation some of the SiC particles are partially or totally detached from the machined surface, and left behind cavities of various sizes and shapes. Some of the detached particles may have passed underneath the tool and were dragged along the surface for a distance and hence resulted in grooves of various lengths. Once again as expected feed marks of the cutting are responsible for the equally spaced long grooves parallel to the direction of cutting velocity. Figure 2.4 is a scanning electron micrograph of the machined surface showing cavity and crushed SiC particles. It has also been observed that the presence of hard particles of SiC in a soft matrix such as aluminum can lead to voids around the silicon carbide particles during the machining operation [3].

Optical and scanning electron microscopy is also used to examine subsurface of the machined components. Figure 2.5 is a scanning electron image of the side surface of the machined surface. From the figure it can be seen that machined surface is irregular. The presence of a micro-crack that initiated at the surface and extended beneath the machined surface is also visible.

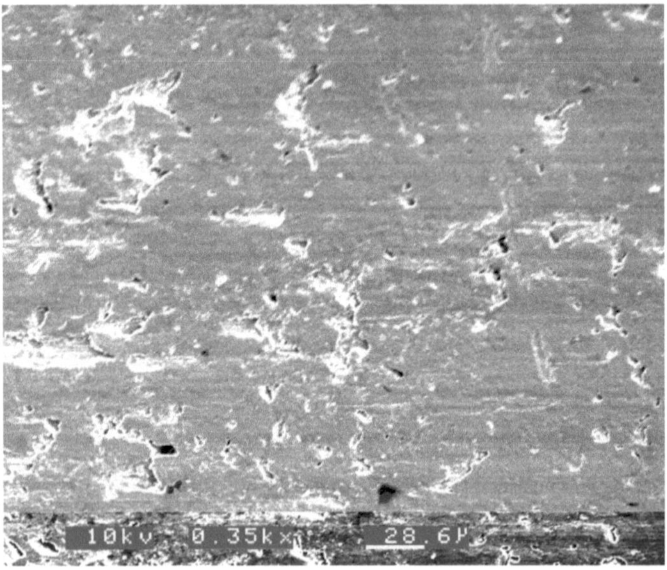

Fig. 2.3 Typical scanning electron micrograph of machined surface, hot-rolled Al/SiC, 20% vol. SiC

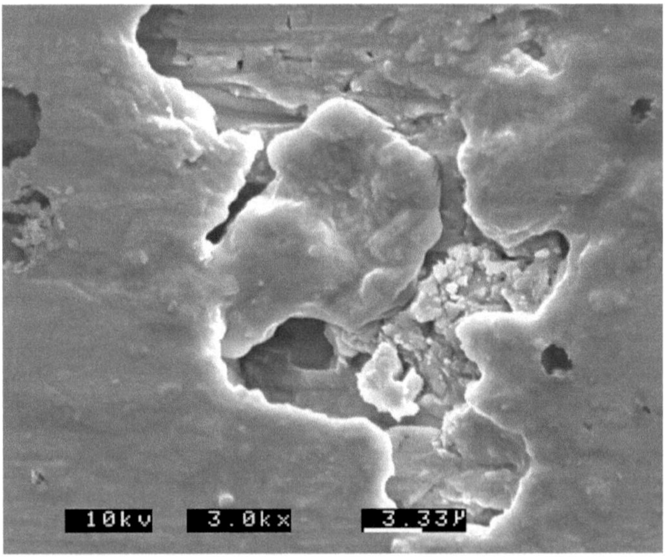

Fig. 2.4 Scanning electron micrograph of machined surface, hot-rolled Al/SiC, 20% vol. SiC

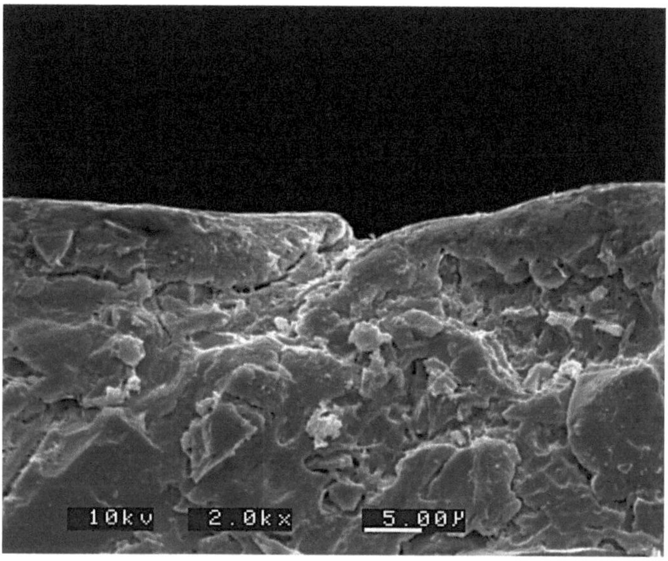

Fig. 2.5 Scanning electron image of the machined subsurface, hot-rolled Al/SiC, 30% vol. SiC

2.3.2 Plastic Deformation Analysis of Surface and Subsurface

Surface region plastic deformation is normally assessed by examining the side surface of the machined specimen that is perpendicular to the machined surface. For this purpose small segments of the test specimen is mounted metallographically in the usual way and then polished and etched to reveal the grain boundaries. Plastic deformation is normally detected from the rotation of grains in the direction of cutting velocity that is visible with the aid of an optical microscope. Macrocracks if present and surface irregularities are also easily detected and observed. Similar observations can be made using scanning electron microscopy. Plastic deformation resulting from the dislocation pile-ups near the machined surface region is detectable using a transmission electron microscope [3]. Microhardness measurements of the surface region (surface and subsurface) are also used to assess subsurface plastic deformation. One of the difficulties in obtaining reliable results is the presence of hard particles of SiC that the indenter may come in contact with, hence will give false reading of the matrix hardness. However, microhardness indentation has been used successfully in the past to evaluate subsurface plastic deformation of machined Al/SiC composites [3].

2.3.3 Residual Stress Analysis of Surface and Subsurface

Residual stress measuring techniques are in general, classified as destructive and nondestructive. The destructive methods such as hole-drilling and deflection-etching involve the destruction of the test samples when evaluating residual stresses. The nondestructive methods that include optical, ultrasonic, electromagnetic, and x-ray diffraction are based on measuring the changes in some physical properties that are caused by the presence of residual stresses. A summary of these techniques is presented in Ref. [4]. Among the nondestructive methods x-ray diffraction may be the most common technique used for measuring residual stresses caused when machining metal matrix composites.

The deflection-etching technique is based on removal of thin layers of stressed material from the machied surface region by electrochemical action. The author of this article has used this technique with great success and hence, it will be discussed further in this section. At this point it will be advisable to know the source of residual stresses in machining.

The source of residual stresses in machining is complex and may be attributed to inhomogeneous plastic deformation caused by mechanical and thermal events associated with the process of chip formation, and the interaction between the tool cutting edge and freshly-machined workpiece surface. Figure 2.6 is a schematic sketch of the residual stress sources and residual stress distribution. The aforementioned sources and their effects are explained as follows. The cutting action of the tool cutting edge and the rubbing or burnishing effect of the tool nose-workpiece contact area may be the leading cause of residual stresses due to mechanical deformation. During the cutting operation, the material ahead of the cutting point experiences compressive plastic deformation and the material behind experiences tensile plastic deformation. If the tensile deformation is more than compressive deformation, the resulting stress is compressive and vice versa [5]. The rubbing and burnishing effect which is similar to surface rolling or shot peening, produces compressive residual stresses. The heating of the surface produces compressive plastic deformation by thermal stresses, then tensile stresses upon cooling. The final state of residual stress distribution in the surface region is the combined effect of the three components [6, 7].

2.3.3.1 Deflection-Etching Technique

Deflection-etching technique was first developed in 1951 [8]. The description of the process and the principle involved is given here. Let us consider a straight metallic bar of rectangular cross-section workpiece having a length of l, a width of w, and a thickness of t. The workpiece is then mounted and clamped and the thickness of the workpiece is reduced by removing a layer of the material using a machining process such as milling, broaching, planing, and grinding. As a consequence of machining residual stresses will be introduced in the surface region of the workpiece that will cause the bar to curve once the bar is unrestrained and removed from the clamping device. The bar may curve concave upward or concave downward (when viewed from the stressed layer) depending on the sign of combined effect of the induced residual stresses as explained in Fig. 2.6. Now if a thin layer of the machined surface is removed by chemical etching this will lead to a partial removal of the stresses that consequently will result in a change in the

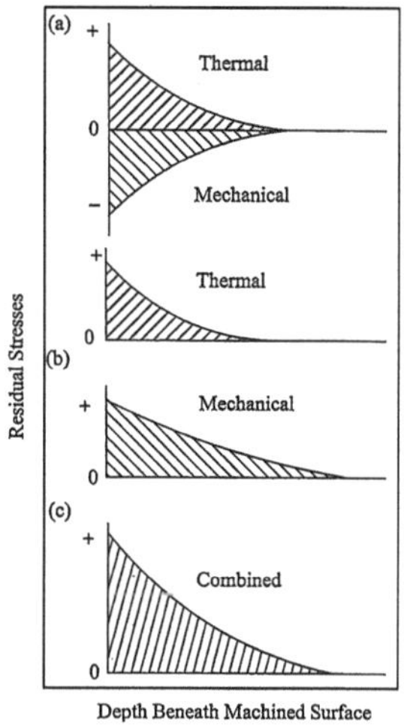

Fig. 2.6 Schematic of residual stress distribution. **a** Rubbing action. **b** Cutting action. **c** Combined effects [6]

radius of curvature of the bar. In this process the remaining residual stresses will be redistributed and a new equilibrium will be established. Theory of elasticity can then be used to determine the stress distribution from the knowledge of the changes in curvature and thickness of the bar due to successive removal of thin layers. A complete analysis of residual stresses for a bar of rectangular cross-section is presented in Ref. [8].

The author of the present article has used a modified deflection-etching technique in evaluating the residual stress distribution in turning operation of ring-shaped metal matrix composites workpieces. A detailed description of the process and the method used in determining residual stress distribution is given elsewhere [9], however, the plot of a typical result is shown in Fig. 2.7. From the figure it can be seen that residual stresses are compressive for the cutting condition and tool geometry used when turning T6-heat treated Al/SiC 20% volume SiC composites test samples. The residual stresses are low at the surface and increase with depth beneath the machined surface. The low residual stresses at the surface may be attributed to relief of stresses due to surface cracking and/or partial or total detachment of hard particles of SiC from the surface as explained previously.

2.3.4 *Surface Topography*

Laser or stylus type profilometers are normally used to measure surface roughness and to obtain surface profile of the machined surface. The presence of hard particles of SiC in metal matrix composites results in higher surface roughness values when compared with surface roughness of matrix metal using optimum and same cutting conditions. This has been demonstrated by mounting samples of 6061 Al matrix metal and 6061Al/SiC composites having 20 and 30% volume SiC

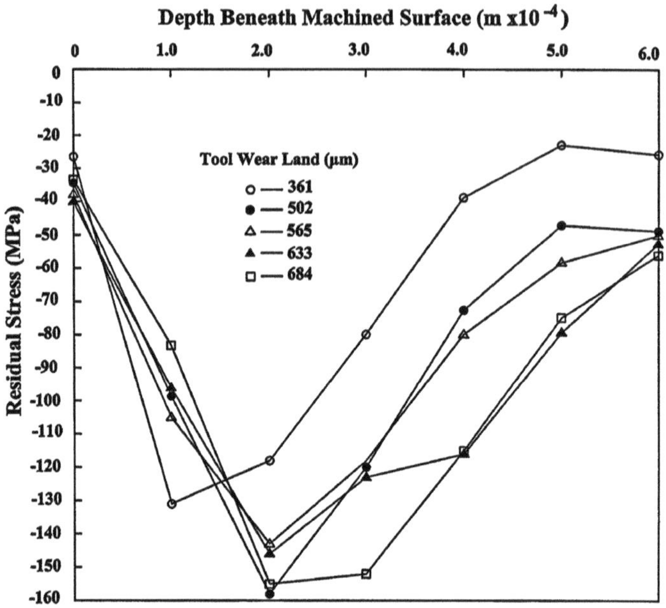

Fig. 2.7 Residual stress distribution—20% vol. SiC—T6 heat treat

reinforcement on the same mounting and then grinding and polishing metallographically in the usual way. The surface roughness of each sample was then measured using a stylus type profilometer and were 0.05, 0.148, and 0.30 μm, respectively, for 6061 Al, 20% SiC, and 30% volume SiC particulates composites. As discussed earlier during machining operation SiC particles are partially or totally removed from the surface and leaving behind cavities of various shapes and depths. Some of these particles are passed underneath the tool flank and dragged by the tool flank along the surface that leads to grooves of various widths and lengths. Therefore, it may not be realistic to anticipate a high-quality surface finish when machining metal matrix composites as it is possible when machining the matrix metal.

2.4 Conclusion

Machining metal matrix composites using traditional cutting methods lead to the damage of surface and subsurface. In general, the severity of the damage is reduced and surface integrity is significantly improved when the cutting speed is increased and the depth of cut and feed rate is decreased. It has been shown that the volume fraction of the reinforcement particles has a great influence on the quality of the machined surface region. The quality of the machined surface improves as the volume fraction of the reinforcement particles is reduced.

Similarly better surface quality can be achieved when the size of the reinforcement particles are reduced. The application of a lubricant does not affect significantly the surface integrity of the machined metal matrix composites. It has also been reported that better surface finish is achieved with whisker reinforced composites as compared with particulate reinforcement composites [10].

References

1. Field M, Kahles JF (1964) The surface integrity of machined and ground high strength steel. DMIC Report 210:54–77
2. Field M, Kahles JF, Cammett JT (1972) A review of measuring methods for surface integrity. Ann CIRP 21(2):219–238
3. El-Gallab M, Sklad M (1997) Machining of Al/SiC particulate metal matrix composites. Part II: Workpiece surface integrity. J Mater Process Technol 83(1–3):277–285
4. Sadat AB, Fahmy AA (1996) Evaluation of residual stresses caused by machining operation. In: Proceeding of the tribology symposium, the energy-sources technology conference and exhibition, sponsored by the ASME Petroleum Division, Houston, TX, pp 389–398
5. Liu RC, Barash MM (1982) Variables governing patterns of mechanical residual stress in a machined surface. Trans ASME J Eng Ind 104:257–264
6. Colwell LV, Sinnot MJ, Tobin JC (1955) The determination of residual stresses in hardened, ground steel. Trans ASME J Eng Ind 77:1099–1104
7. Hannover BE, Cincinnati CTJ, Aachen KW, Ljubljana PL, Leuven LP, Hannover HKT (1982) Residual stresses—measurement and causes in machining operation. CIRP Ann Manuf Technol 31(2):491–510
8. Frisch J, Thomsen EG (1951) Residual grinding stresses in mild steel. Trans ASME 73: 337–345
9. Sadat AB, Bailey JA (1985) Residual stress distribution in machining an annealed bearing bronze. Int J Mech Sci 27(11):717 721
10. Cheung CF, Chan KC, To S, Lee WB (2002) Effect of reinforcement in ultra-precision machining of AL6061/SiC metal matrix composites. Scripta Mater 47:77–82

Chapter 3
Machinability Aspects of Metal Matrix Composites

Antoniomaria Di Ilio and Alfonso Paoletti

Machining of Metal Matrix Composites (MMCs) is notoriously known to be difficult due to both the presence of two or more distinct phases, one of which is very abrasive, and for the marked differences between the two constituents: the hard ceramic reinforcement and the ductile metal matrix. For this reason, a number of efforts have been made to produce metal matrix composite components in near-net-shape forms. However, such parts always have to be machined to match the final design requirements.

The aim of this chapter is to give an overview of the present knowledge about the machinability of MMCs, which represents one of the most important concerns which tends to limit the number of applications of these materials in industry. After an introduction about the meaning of machinability, the main characteristics of the material, which can play a significant role on the machinability of MMCs, are presented. Such characteristics are then analysed and discussed in the subsequent sections as regards their influence on cutting tool wear, surface integrity, cutting forces and chip formation.

3.1 Introduction

Machinability is not a peculiar characteristic of the material univocally definable and measurable. Generally, it is considered as the attitude of the material to be machined, i.e. the level of difficulty encountered in the shaping a workpiece employing a tool. For example, a medium carbon steel is recognised to be easier to machine with respect to a heat-resistant alloy, grey sand cast iron exhibits less difficulty than the same material poured in die, while a low-carbon steel can show more problems in

A. Di Ilio (✉) · A. Paoletti
Università dell'Aquila, Piazza V. Rivera 1, 67100 L'Aquila, Italy
e-mail: antoniomaria.diilio@univaq.it

J. P. Davim (ed.), *Machining of Metal Matrix Composites*,
DOI: 10.1007/978-0-85729-938-3_3, © Springer-Verlag London Limited 2012

machining with respect to some alloy steels. Therefore, due to different methods and operating conditions, as well as the tool characteristics, the concept of machinability is often equivocal and not easily measurable and comparable.

A comparison of all the characteristics (physical, mechanical and thermal) of the material, which can influence the machining process, could allow a deeper understanding about the machinability in a particular situation. However, such characteristics are often difficult to be obtained by material manufacturers.

Chemical and mechanical properties of the constituent materials, structural properties, which in turn are greatly influenced by thermal treatments, reinforcing particles and other inclusions, surface alterations, such as oxidation or hardened layers resulting from a previous manufacturing process, all affect the machinability. However, also the built-up edge (BUE) which, for a given material, depends on the machining conditions, tool geometry, type of machine tool, as well as the general machining conditions determines the material machinability.

A complete and correct approach to study the machinability has to take into account tool performance, workpiece surface integrity, power required for material removal, as well as type and shape of the chips, which can represent a serious problem to draw the removed material away from the cutting zone.

Depending on the aim of the particular machining operation, machinability of MMCs can be assessed by taking into consideration different machining parameters. These parameters can lead to machinability ranks which can differ very much between each other.

3.2 Parameters of Machinability

In general, machinability can be described, from an operating point of view, by means of one or more than one of the following factors:

- cutting tool wear and tool life,
- surface and sub-surface integrity,
- cutting forces and power,
- tendency to form BUE,
- chip formation.

Through a proper set of information concerning the material properties as well as suitable machining tests it is possible to evaluate the machinability of the material, in order to satisfy both specific and general cutting conditions.

However, what can be considered a good machinability in one particular situation, might not be as good as it is in another.

For example, bulk strength and hardness of the material are not representative indicators of the machinability without considering the hard and abrasive particles, as well as other further important factors such as additives, microstructure of the matrix, tendency of the material to stick on the tool face, which can greatly affect the machinability.

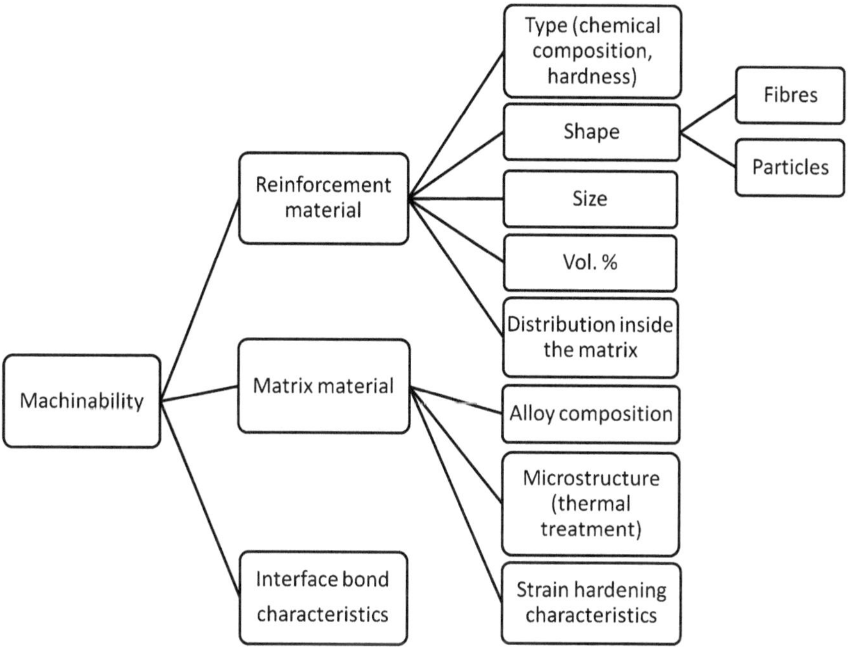

Fig. 3.1 Characteristics of the constituent materials related to machinability of MMCs

In Fig. 3.1, the characteristics of the constituent materials which can affect the machinability of MMCs are reported. As it can be deduced from the block scheme, the knowledge of the properties of the single components is not sufficient to forecast the behaviour of the material in the cutting process. It is indeed important to consider the interactions between the two constituents, in particular the strength of the bond at the interface, which can affect both the quality of the machined surface and the tool wear, as well as all those characteristics which descend from the previous manufacturing process employed to produce the composite, e.g. the homogeneity of reinforcement distribution, its alignment in one direction, the crystalline structure of the matrix and so on.

In the next sections, these characteristics will be discussed at the light of the results of investigations reported in the literature.

3.3 Cutting Tool Wear and Tool Life

3.3.1 Tool Wear

The main wear mechanism when machining MMCs is abrasive wear. The predominant wear develops principally at the flank face of the tool, but abrasion marks are also present on the rake face.

Attrition wear, also called adhesion wear, which can be associated with the formation of BUE, has often been detected by researchers.

Crater wear may also be present, especially at the highest speed, though it does not represent the determinant cause of tool failure [1].

Detailed examinations of the cutting edges under scanning electron microscope (SEM), revealed that edge chipping can occur when machining Al-SiC_p using uncoated cemented carbide tools, particularly at high speed, as reported in [2]. The same authors noted that at high-cutting speeds, due to edge chipping and uncoated carbides can produce surface finish poorer than the coated ones, which is an opposite behaviour to what generally occurs when machining homogeneous materials such as aluminium alloys.

Owing to the extremely abrasive nature of MMCs, only polycrystalline diamond Polycrystalline Diamond (PCD) brazed tools reach an acceptable life [3–7], since diamond is harder than silicon carbide (SiC) and alumina (Al_2O_3) and does not exhibits any chemical tendency to react with the workpiece. Therefore, many researchers recommend the use of such a material in machining MMCs. As a general rule, HSS tools must be excluded, while cemented carbide tools, both coated and uncoated, and even TiN-coated high-speed steels can be an economic choice for short-run productions [2, 8]. Some results indicate that the coating is not economically advantageous, due to the small increase in the tool life compared with the relatively high cost [1, 9].

As far as tool life assessment is concerned, they go from the simple Taylor's relationship, which takes into account a prefixed acceptable flank wear, to more sophisticated empirical models developed either to obtain a better prediction or a more economical way to determine the relative parameters experimentally.

3.3.1.1 Influence of the Reinforcement Material

The most popular reinforcement materials are SiC and Al_2O_3, though B_4C and TiC are also used in some cases. All these materials are extremely hard and extremely abrasive for the tools. Comparative investigations on the machinability of MMCs containing different reinforcement materials are not many. The major part of papers concerning the machining of MMCs deals with composites reinforced with SiC, both in form of whiskers or particles.

A comparative analysis between the different reinforcement materials have been made by Ray et al. [10]. They analysed the machinability of in situ fabricated aluminium-matrix MMCs reinforced with TiC, $TiAl_3$ and Si; the behaviour of the composites were compared to that of non-reinforced aluminium. They found that the best machinability in terms of surface roughness of the workpiece and cutting forces were exhibited when machining Al-TiC MMCs. For all the composites, the cutting force increased with the reinforcement content; Al-TiC composite showed the lowest one and a reduced roughness. This fact was attributed to the absence of BUE for Al-TiC, which gives lesser tool attrition wear and a consequently better surface finish.

As far as the comparison between SiC and Al_2O_3 is concerned, it should be considered that the former is harder than the latter, therefore one should expect that tool wear be faster when machining MMCs containing SiC particles.

Few investigations deals with the machinability of aluminium-matrix composites reinforced with B_4C. Karacas et al. [11] performed an experimental investigation aimed to compare the performances of coated and uncoated tungsten carbides in machining such a composite. The authors report the trends of flank wear and concluded that the best performance was exhibited by triple coated carbide tools. They found that besides abrasive wear, also adhesive wear associated to BUE formation, which occurred at all cutting speeds, was significant.

The machinability of MMCs also depends on the interfacial bond which can notably influence the quality of the machining in terms of surface roughness and material damage. The strength of the interfacial bond depends on both the nature of the reinforcement and the matrix composition.

As far as the quantity of the reinforcement is concerned, it is generally accepted that machinability reduces as the volume fraction increases. Cheung et al. [12] found that the surface roughness increased with increasing the volume fraction of whiskers. Larger tool marks and surface waviness formed on the machined surface as volume fraction increased. This result can be explained with the following mechanism: after the tool passes, the stress relaxation of the tool cutting edge associated to cutting of hard SiC whiskers induces vibrations provoking an increase of the surface waviness.

Chandrasekaran and Johansson [13] investigated the effect of SiC particle size and concluded that there exists an optimum value of reinforcement content to which corresponds an optimum value of feed rate.

Xiaoping and Seah [14] found that both the percentage of reinforcement and the size of particles are important for the machinability in terms of tool life. In particular, they found that there exists a critical value of SiC wt% above which the flank wear rate increases rapidly. Likewise, the flank wear rate shows an accelerated increase above a critical value of the particle size; moreover, the critical reinforcement content decreases as the particle size increases.

Ciftci et al. [2] also investigated the influence of the reinforcement particle size on the performance of Cubic Boron Nitride (CBN) tools in turning. They found that MMCs containing large SiC particles (110 μm) were unsuitable to be cut with CBN tools due to the heavy fractures of the cutting edge.

3.3.1.2 Influence of the Matrix Material

The most popular matrix material employed in MMCs is aluminium, while other matrices, such as magnesium and titanium are employed occasionally.

Only few investigations deal with the machining of MMCs with magnesium matrix [9, 15]. Although most of the challenges in machining Mg MMCs are the same as those for aluminium-based MMCs, there are specific metal matrix concerns that need to be considered. Magnesium provides a better bond for the

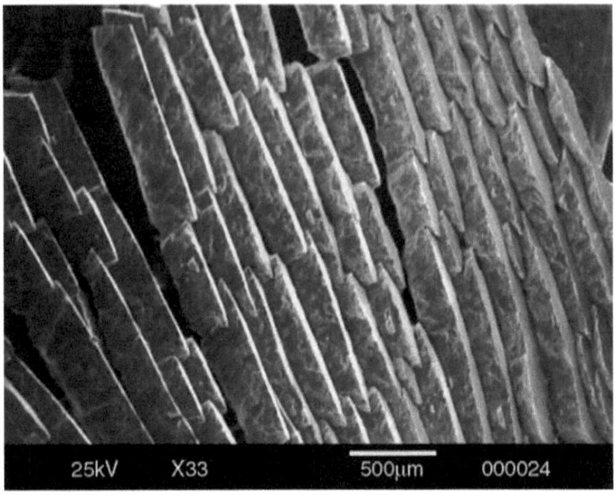

Fig. 3.2 SEM photograph of the free surface side of a chip obtained in finish turning of Mg-SiC_p 20 vol% MMC showing extreme segmentation [9]

reinforcement particles, which greatly reduces debonding under the tool action. This fact gives a higher surface strength which is important in applications were fatigue resistance of the component can be a concern. Magnesium has been sometimes added to aluminium-matrix composites in order to promote the wetting of the particles during the fabrication of the composite in order to obtain an increase of the bond strength [16].

Some interesting results can be found in the experimental work carried out by Pederson and Ramulu [9], where the finish machining characteristics of a silicon carbide particle-reinforced magnesium were instigated. They found that the surface finish of the machined part was excellent (Ra = 0.2–0.3 μm) and very near to the theoretical one and, in addition, almost independent of the depth of cut side angle and cutting speed. The cutting forces were rather constant over a wide range of cutting speeds, indicating a low tendency to form BUE. The obtained chips exhibited basically the same highly-segmented nature, the free surface side of them evidencing an extreme segmentation (Fig. 3.2) which justifies the regular trend of the cutting force during machining. In addition they were particularly fragile and easily broken upon collection.

Chambers [15] underlined that the role of the matrix can be important in determining machinability more than one can expect. From his experiments he found that a softer matrix can cause higher tool wear. In fact, soft matrix is preferentially worn, thus leaving the reinforcement material exposed and free to abrade the cutting tool.

Machinability of aluminium-based MMCs can be increased by incorporating small quantity of graphite into the aluminium matrix [17, 18]. The results of the investigation evidence that the inclusion of graphite is effective in reducing the thrust force in drilling and the burr height at the exit side; in addition the presence of graphite helps in forming discontinuous chips, which is particularly important in drilling for an easier removal of them from the hole being drilled. However, the surface roughness increases due to graphite pull-out. The reduction of machining forces is attributable both to the solid lubricating property of the graphite fibres, which reduces the friction between the chip and the tool, and to the reduction in shear flow stress of the material, as also observed by Brown and Surappa [19].

The heat treatment of the matrix shows a significant influence on the machinability: higher matrix hardness, as a result of heat treatment, generally shortens the tool life [20]. Barnes et al. [21] suggest that the best material machinability in terms of tool life is the soft condition of the matrix; in practise, they suggest to machine the material before any thermal treatment of solubilization or ageing. However, the increase of tool life does not eliminate the wear, while, on the other hand, further problems become more important with respect to a harder matrix. In fact, although the material in the softer condition gives a lower tool wear, lower forces and lower subsurface damage, it should be noted that it produces the largest burrs. This fact could represent a significant problem related to the need of removing, with additional work time, a hardened material sometimes located in difficult to reach zones.

3.4 Surface and Sub-Surface Integrity

From the theory of metal cutting, the analysis of surface integrity is believed to be one of the most effective ways for understanding the machining characteristics of a material.

Material removal by means of a tool is invariably associated to some damage of the machined workpiece, which can be described as surface and subsurface integrity. One important quality parameter is the surface roughness which, as known, depends on the tool nose radius and feed speed (ideal surface) and, in addition, on how the separation of the chip from the machined surface occurs, both giving rise to the so-called real surface. The second aspect, which concurs to the formation of the real surface, is particularly important and complex for MMCs, due to the high heterogeneity and anisotropy at micro-geometry level of such materials.

Another aspect dealing with the quality of the machining result is represented by microstructural alterations the material undergoes on account of plastic deformation and temperature rise in the layer which extends up to a certain depth from the outer surface. These alterations can be burnings, strain hardening, different grain sizes and residual stresses, all of them being potentially influent on the mechanical, chemical, electrical and other properties of the surface, especially long-term ones such as fatigue and stress corrosion.

Due to the high thermal conductivity of MMCs, as they are generally based on aluminium matrix, damage such as burning associated to temperature rise is not reported in the literature, however, grains undergo high deformation under the tool action with consequent hardening and residual stresses rising. This process is particularly complex in the case of MMCs compared to non-reinforced alloys due to the presence of particles or whiskers of reinforcement.

In the following, further considerations will be made concerning the aspects of machinability, related to surface integrity, such as surface finish, microstructural alterations and residual stresses.

3.4.1 Surface Finish

The surface finish obtainable with MMCs is a matter where opinions are often divided. This occurs because the results obtained by different investigators refer to different materials tested, as well as different working conditions. In fact, the surface quality depends on the shape, the size and the volume fraction of the reinforcement in addition to the type of matrix and its conditions.

The machined surface generally exhibits good surface finish, which can be attributed to the grinding effect produced by the reinforcement particles [22, 23].

According to Lin et al. [24], MMCs are difficult to be machined to a good surface finish, depending on the fact that the particles or whiskers are truncated at the surface level or uproot out of the matrix giving rise to pits and scratches.

Therefore, the surface quality seems to be strongly related to shape, size, volume, as well as the distribution of the reinforcement elements.

Cheung et al. [12] carried out an analysis on the mechanisms of surface generation in turning Al-SiC composites with diamond tools, where the only variable were the shape of reinforcement, i.e. powder or whiskers. The authors found that two types of material removal mechanisms exist: the cut-through mechanism and the pull-out mechanism, the former giving better surface finish, the latter leaving cracks and pits on the machined surface causing poor surface finish. They found that the through cut mechanism invariably occurred when machining whiskers reinforced composites, independently of the cutting direction with respect to the lying of the fibres. This behaviour can be explained, according to the authors, with the higher effective interface area between fibres and matrix with respect to powder.

The effect of volume fraction is such that to improve the surface finish as the number of fibres decreases. This behaviour was explained through the vibrations arising due to stress relaxation whenever a fibre is cut, the triggering of such events being less likelihood as the volume fraction of reinforcement decreases.

As far as the influence of machining parameters is concerned, the surface finish generally worsens with increasing feed rate, but it does not change significantly when varying the cutting speed.

3.4.2 Sub-Surface Damage

Microstructure alterations consisting of strain hardening, interfacial debonding of the reinforcement and particles alignment occur due to the plastic deformation induced by the cutting tool.

El-Gallab and Sklad [22] found that plastic deformation due to machining extends up to 100 μm beneath the machined surface, giving rise to an increase of the material hardness. The authors found that, close to the machine surface, microhardness values generally exhibit an increasing trend starting from values slightly higher than those of the basic material up to a maximum, then decrease up to reach the original value (Fig. 3.3).

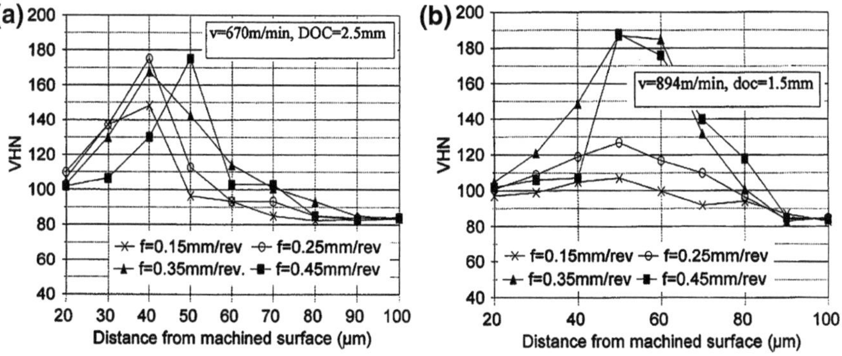

Fig. 3.3 Microhardness profiles as a function of depth below the machined surface obtained in turning Al-SiCp 20 vol% [22]

The cause of such a trend is due, according to the authors, to dislocation pile-up, formation of finer grains close to the machined surface and precipitation of silicates needle-like particles along the maximum shear planes.

Dandekar and Shin [25] introduced a 3D cohesive model for studying the damage development during the machining of Al-SiC composites. In this study the authors used the results of a previous study carried out by Zhang et al. [26], who characterized the interfacial debonding computationally for an aluminium-matrix composite reinforced with boron carbide particles (Al-B_4C). On the basis of the assumption of a similar behavior for both Al-SiC and B_4C composites, they estimated the energy separation for Al-SiC. They found that the damage depth, in terms of particle debonding, is a primary function of the feed rate, which corresponds to higher cutting forces (Fig. 3.4). The cutting speed also produces an increase of damage depth, though its influence is minimal.

Hung et al. [20] found that CBN and PCD tools induce minimum damage to the subsurface of the material and then these materials are particularly suitably for finishing operations. This result can be attributed to the high-thermal conductivity of such materials and to the very small edge radius able to withstand enough to the abrasive action of the hard particles.

3.4.3 Residual Stresses

Residual stresses distribution represents an important parameter of a machined component surface integrity which is important for its fatigue strength. These stresses depend either on the work material or on the machining parameters.

Only few studies devoted to this aspect have been reported up to date, and most of them suggest that both mechanical and thermal effects are involved. Due to the non-homogeneous structure of MMCs, the effects of machining parameters on

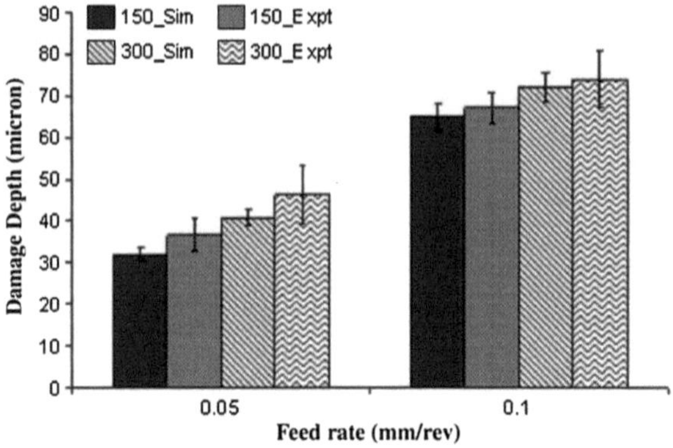

Fig. 3.4 Theoretical and experimental results of damage depth when machining Al-SiC 20 vol% at different depths of cut and cutting speeds [25]

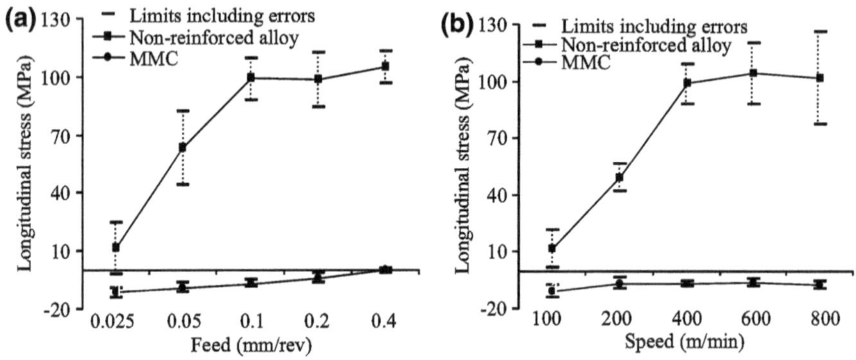

Fig. 3.5 Effect of **a** feed (at speed 400 m/min), and **b** speed (at feed 0.1 mm/rev) on residual stress along longitudinal direction [27]

surface residual stress are quite different with respect to the conventional homogeneous materials, as shown by Pramanik et al. [27] in turning an Al-SiC_p 20 vol%. Their results evidence that the longitudinal surface-residual stress parallel to the machined surface, for the non-reinforced alloy, increases with both feed rate and cutting speed (Fig. 3.5). On the contrary, MMCs show slightly-compressive stresses which, in addition, seem to be almost independent of cutting parameters; similar results are reported for the transverse residual stresses.

According to Capello [28], three mechanisms are involved in the rising of residual stresses, namely mechanical (plastic deformation), thermal (thermal plastic flow) and physical (specific volume variation). Tensile residual stresses are caused by thermal effects, while compressive stresses are caused by mechanical deformation induced by the tool action. The mechanism at the basis of the behaviour of MMCs is the result of the opposite contribution to the residual stresses of the thermal mechanism and the mechanical effects generated by the presence of reinforcement particles. Pramanik et al. [27] suggested three factors that give their contribution to this phenomenon, i.e. (a) restriction of matrix flow (b) indentation effect of the particles and (c) compression of the matrix between the particles and the tool. Mechanisms (b) and (c) can be explained as follows: when the tool passes, the particles are elastically forced inside the beneath material, the subsequent springback of the particles towards the outer surface, after the tool pass, produces a compressive in-plane stress in the outer layer. This effect increases with either the growth of the elastic compression, which occurs with the increasing of the undeformed chip thickness and the matrix softening in the outer layer. The latter occurs with the increase in cutting speed, thus compensating the tensile stresses generated by the thermal plastic flow which is predominant at higher speeds.

3.5 Cutting Forces

Cutting forces represent an important factor of machinability evaluation: higher cutting forces means higher stresses on the tool, causing wear and vibrations, and higher stresses on the material causing more damage at higher depth from the machined surface. Moreover, higher forces means that a larger energy is transformed into heat, thus giving rise to higher temperature of both the tool and the workpiece surface. The damage can consist of strain hardening, structural variations, debonding as well as residual stresses. In finish machining, cutting forces should be kept low in order to minimise such damage, as well as to achieve better surface finish and tolerances.

Cutting forces depend on both the material properties and machining parameters.

According to Morin et al. [29], cutting forces are maily controlled by the matrix and not by the particles, however. pressure peaks on the cutting edge are notably influenced by the type and content of the reinfocement and by the bond strength at the interface.

It is a common practise to estimate the cutting forces multiplying the specific cutting force by the undeformed chip section. The specific cutting force depends not only upon the material, but also on the cutting speed and chip thickness, which in turn depends on the feed rate and the inclination angle of the cutting edge. Since many authors have found that lower abrasive tool wear occurs at high feed rate and low cutting speeds, it follows that a reasonable compromise should be adopted in order to allow reduced tool wear and low cutting forces.

Cutting forces also depend on the coefficient of friction between the tool and the material, therefore all conditions promoting the rising of BUE lead to an increase of the cutting forces. Unfortunately, this tendency is particularly high for MMCs, and the remedy consists in adopting high cutting speeds, though it has been noted that also at the highest speeds BUE cannot be totally eliminated.

The use of cutting fluids is not always recommended, since they can give greater tool wear by helping to maintain in contact the abrasive particles to the cutting tool [30].

As stated above, the use of small quantity of graphite inserted into the matrix material during the fabrication of the composite, while it does not reduce the composite strength appreciably, it allows to reduce the cutting force through the reduction of the coefficient of friction and the shear strength.

Since cutting forces grow as the tool wear increases, due to the larger contact area against the workpiece, the quality of the machining, in terms of tolerances, seriously reduces if the tool material does not have a sufficient abrasion resistance.

3.6 Chip Formation

For a good machinability, it is desirable to have continuous chips in short segments without the use of chip breakers; the importance of this fact has been recognized by Armarego and Brown [31] already long ago. Even minor changes in the chip-formation process can cause degradation of surface finish, poor dimensional accuracy as well as shorter tool life. Therefore, it is important to have short chips without employing tool breaker in order to have a continuous machining operation. Chip formation, and the related influence on the material removal from the cutting zone, is considered a crucial factor for increasing the number of applications of MMCs in industry.

Chip formation is influenced by ductility, thermal conductivity and microstructure; however, physical phenomena, such as instability in the cutting process, e.g. the variation in the inclination angle, can change the chip-formation mechanism.

Thermal conductivity influences both the thickness of the material undergoing temperature increase and mean temperature. In fact, as the thermal conductivity increases a thicker layer of material undergoes warming, while the mean temperature of the cutting zone decreases. This fact produces a change in the chip formation which tends to be more fragmented due to the more brittleness which the material exhibits at lower temperature.

Cheung et al. [12] studied the influence of SiC on the chip-formation process for an aluminium alloy reinforced with SiC particles by employing the quick-stop test in turning operation. They concluded that the reduction of the ductility of the material, achieved by the addiction of SiC particles, favours the production of semi-continuous chips. The cracks which bring to fragmentation of the chip initiate at the outer surface of the chip because of the shear stress caused by the tool rake face; other small voids form after debonding of the particles from the matrix, caused by stress concentration at the edge of the particles, which subsequently coalesce to form a fracture along the shear plane.

By observing the chips produced in machining Al-Al_2O_3 MMCs, it has been noted that the particles tend to pile-up along the shear plane dividing the chip in small

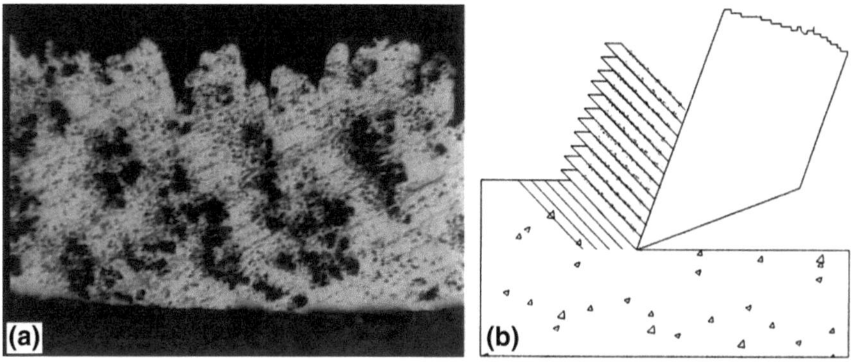

Fig. 3.6 Concentration of reinforcing particles along the shear planes: **a** $v_c = 1{,}000$ m/min, $f = 0.03$ mm/rev; **b** pile-up formation according to the Pijspanen model [1]

fragments [1]. This phenomenon can further explain why MMCs tend to give discontinuous chips in a wide range of cutting speeds. In fact, the phenomenon has been found to be more evident as the cutting speed or feed increases due to the increased temperature which allows the particles to move more freely through the matrix. The model proposed by the authors is depicted in Fig. 3.6.

Similar results have been obtained also by other authors when machining Al-SiC composites both in turning [22, 32] and in drilling [33].

The crack formation is also influenced by the tensile residual stresses acting on the ductile matrix. These stresses are generated in the manufacturing process of the composite during cooling from the process temperature to the ambient one, as a consequence of the thermal coefficient mismatch between the ceramic particles and the surrounding material [34]. Due to the larger coefficient of thermal expansion of the aluminium matrix with respect to that of the ceramic reinforcement, the residual stresses are tensile in the matrix and compressive in the reinforcement particles.

3.7 Conclusion

The evaluation of machinability of MMCs is a very complicated issue for the large number of the material characteristics involved.

The main problem, which is crucial indeed for the increase of industrial applications of MMCs, is represented by the tool wear caused by the hard particles or whiskers of reinforcement. For this reason tool materials able to give acceptable tool life are PCD or CBN; only small batches of workpieces can be processed with coated and even uncoated carbides.

Surface finish is generally good and better with respect to that obtained when machining the matrix alloy alone, but the results may worsen under particular conditions related to the type of composite, cutting tool conditions and machining parameters. Material damage due to thermal and mechanical effects of machining, as well as the residual stresses, is generally small and negligible for the most part of applications.

The chips obtained from the machining process are generally short and easily removable from the cutting zone.

Acknowledgments The authors wish to acknowledge Elsevier Publishers for their kind permission to reuse figures and illustrations from a previously-published material.

References

1. Iuliano L, Settineri L, Gatto A (1998) High-speed turning experiments on metal matrix composites. Composites Part A 29:1501–1509
2. Ciftci I, Turker M, Seker U (2004) Evaluation of tool wear when machining SiCp-reinforced Al-2014 alloy matrix composites. Mater Des 25:251–255
3. Lane G (1992) The effect of different reinforcement on PCDtool life for aluminuim composites. In: Proceedings of the Machining of Composite Materials Symposium. ASM Materials Week, Chicago, IL, pp 3–15
4. Lin JT, Bhattacharya D, Lane C (1995) Machinability of a silicon carbide reinforced aluminium metal matrix composite. Wear 181–183:883–888
5. Davim JP, Baptista AM (2000) Relationship betwen cutting force and PDC cutting tool wear in machining silicon carbide reinforced aluminium. J Mater Process Technol 103:417–423
6. Davim JP, Conceição António CA (2001) Optimal drilling of particulate metal matrix composites based on experimental and numerical procedures. Int J Mach Tools Manuf 41:21–31
7. Paulo DJ (2002) Diamond tool performance in machining metal-matrix composites. J Mater Process Technol 128:100–105
8. Norrul Haq A, Marimuthu P, Jeyapaul R (2008) Multi response optimization of machining parameters of drilling Al/SiC metal matrix composite using grey relational analysis in the Taguci method. Int J Adv Manuf Technol 37:250–255
9. Pederson W, Ramulu M (2006) Facing with carbide tools SiCp/Mg metal matrix composites. J Mater Process Technol 172:417–423
10. Rai RN, Datta GL, Chakraborty M, Chattopadhyay AB (2006) A study on the machinability behaviour of Al-TiC composite prepared by in situ technique. Mater Sci Eng A 428:34–40
11. Karacas MS, Acir A, Ubeyli M, Ogel B (2006) Effect of cutting speed on toolperformance in milling of B4Cp reinfrced aluminium metal matrix composites. J Mater Process Technol 178:241–246
12. Cheung CF, Chan KC, To S, Lee WB (2002) Effect of reinforcement in ultra-precision machining of Al6061/SiC metal matrix composites. Scripta Materialia 47:77–82
13. Chandrasekaran H, Johansson JO (1996) On the behaviour of fibre/particle reinfeorced aluminium alloy matrix composites in milling and grinding. VDI Ber 1276:463–478
14. Xiaoping Li, Seah WKH (2001) Tool wear acceleration in relation to workpiece reinforcement percentage in cutting of metal matrix composites. Wear 247:161–171
15. Chambers AR (1996) The machinability of light alloy MMCs. Composites Part A 27A:143–147
16. Li XP, Lu L (2003) Study of reinforcement percentage of Al-Mg-SiC MMC in relation to the mechanical properties and machinability. Mat Sci Forum 437–438:185–188
17. Basavarajappa S, Chandramohan G, Davim JP, Prabu M, Mokund K, Ashwin M, PrasannaKumar M (2008) Drilling of hybrid aluminium matrix composites. Int J Adv Manuf Technol 35:1244–1250

18. Basavarajappa S, Chandramohan G, Davim JP (2006) Some studies on drilling of hybrid metal matrix composites based on Taguchi techniques. J Mater Process Technol 196:332–338
19. Brown CA, Surappa MK (1988) The machinability of a cast aluminium alloy-graphite particle composite. Mater Sci Eng A 102(1):31–37
20. Hung NP, Boey FYC, Khor KA, Phua YS, Lee HF (1996) Machinability of aluminium alloys reinforced with silicon carbide particulates. J Mater Process Technol 56:966–977
21. Barnes S, Pashby IR, Hashim B (1999) Effect of heat treatment on the drilling performance of Aluminium/SiC MMCs. Appl Compos Mater 6:121–138
22. El-Gallab M, Sklad M (1998) Machining of Al/SiC particulate metal matrix composites. Part II: Workpiece surface integrity. J Mater Process Technol 83:277–286
23. Reddy NSK, Kwan-Sup S, Yang M (2008) Experimental study of surface integrity duruing end milling of Al-SiC particulate metal-matrix composites. J Mater Process Technol 201:574–579
24. Lin JT, Bhattacharya D, Ferguson WG (1998) Chip formation in the machining of SiC-particle-reirforced aluminium-matrix composites. Compos Sci Technol 58:285–291
25. Dandekar CR, Shin YC (2009) Multi-step 3-D finite element modelling of substrate damege in machining particulate reinferced metal matrix composites. Composites Part A 40:1231–1239
26. Zhang H, Ramesh KT, Chin ESC (2005) Effects of interfacial debonding on rate dependent response of metal matrix composites. Acta Mater 53:687–700
27. Pramanik A, Zhang LC, Arsecularatne JA (2008) Machining of metal matrix composites: effect of ceramic particles on residual stress surface roughness and chip formation. Int J Mach Tools Manuf 48:1613–1625
28. Capello E (2005) Residual stress in turning, Part I: influence of process parameters. J Mater Process Technol 160:221–228
29. Morin E, Masounave J, Laufer EE (1995) Effect of drill wear on cutting forces in the drilling of metal-matrix composites. Wear 184:11–16
30. Cronjager WM, Meister D (1992) Machining of fibre and particle reinforced aluminium. Ann CIRP 41(1):63–66
31. Armarego EJA, Brown A (1969) Machining of metals. Englewood Cliffs, Prentice-Hall, pp 36–62
32. Kannan S, Kishawy HA (2008) Tribological aspects of machining aluminium metal matrix composites. J Mater Process Technol 198:399–406
33. Tosun G, Muratoglu M (2004) The drilling of an Al-SiCp metal-matrix composite. Part I: Microstructure. Compos Sci Technol 64:299–308
34. Olivas ER, Swadener JG, Shen YL (2006) Nanoindentation measurement of surface residual stresses in particle-reinforced metal matrix composites. Scripta Materialia 54:263–268

Chapter 4
Traditional Machining Processes of MMC

H. A. Kishawy, S. Kannan and G. Parker

The present chapter focuses on the problems encountered during high-speed machining of Metal Matrix Composites (MMCs) and how it affects tool life, surface quality and integrity and cutting forces generated during different machining processes such as turning, drilling and milling. It should be mentioned here that most studies on the machinability of MMCs have been based entirely on experimental results while very few analytical models have yet been developed.

4.1 Introduction

In recent years, MMCs have received considerable interests among researchers and in several industries due to their high-specific strength, stiffness, and superior wear resistance. These properties, coupled with low density, and ability to operate at elevated temperatures, have made these materials an excellent candidate to manufacture a wide range of products from aerospace parts to sports goods. However, due to the abrasive nature of the reinforcements, MMC is classified as a difficult to cut material. The deterioration of the workpiece surface finish and high

H. A. Kishawy (✉) · G. Parker
Engineering and Applied Science, University of Ontario Institute of Technology,
2000 Simcoe Street North Oshawa, ON L1H 7K4, Canada
e-mail: hossam.kishawy@uoit.ca

G. Parker
e-mail: Grant.Parker@uoit.ca

S. Kannan
Manufacturing Technology Group, Wilmore Road, Rolls-Royce plc,
Derby, DE24 8BJ, UK
e-mail: jeshtasathish@yahoo.co.in

J. P. Davim (ed.), *Machining of Metal Matrix Composites*,
DOI: 10.1007/978-0-85729-938-3_4,

Table 4.1 Typical reinforcements of MMCs [5]

Type of MMCs	Reinforcements
Particle reinforced	Al_2O_3, SiC, WC, TiC, B_4C
Continuous fiber reinforced	Al_2O_3, SiC, B, C, Al_2O_3 + SiO_2, Nb-Ti, Nb_2Sn, Si_2N_4
Whiskers or short fiber	Al_2O_3, B, C, Al_2O_3 + SiO_2,SiC, TiB_2

tool wear rates are some of the major problems that limit the widespread use of MMCs in industry.

MMCs can be categorized into the following three categories according to the types of reinforcements:

1. Particle reinforced MMCs;
2. Whiskers or short-fiber reinforced MMCs;
3. Long-fiber or continuous fiber reinforced MMCs.

The size of a reinforcement particle used in a typical MMC ranges from 2 to 200 μm and the volume fraction varies between 5 and 40%. Also, matrices are generally based on light metal alloys like aluminum, copper, zinc, steel and magnesium. Reinforcements used are generally of abrasive nature with high hardness. Typical examples of reinforcement material include silicon carbide, titanium carbide, aluminum oxide, soda ash, boron nitride and graphite. Table 4.1 shows typical types of reinforcements used in each category of MMC. Generally these composites are fabricated by three techniques namely solid state, liquid state and powder metallurgy. Among them liquid state is the most widely used due to its low-cost and effortless production process. Some new methods like melt-stir casting, continuous casting, direct-chill casting etc., have been reported by several researchers [1–4] for fabrication of MMC.

In the manufacturing industry, most conventional machining processes such as turning, drilling and milling are widely utilized for machining composite materials in which reinforcements (such as glass, graphite, boron, alumina and silicon carbide) are highly abrasive and hard; sometimes as hard as or even harder than the tool material. Although the latest innovative manufacturing processes can produce near-net shape components to minimize machining, final machining and finishing processes are generally still required to fabricate a MMC component to the final dimensions. Although the hard reinforcements in MMCs provide high-wear resistance, they are detrimental to cutting tools and forming dies. High-tool wear and poor surface quality are generally observed during machining with various tools.

4.2 Turning of MMC

Most of the recent research on MMCs are based on light MMC materials. These studies include chip formation, tool life, tool wear mechanisms, machined surface quality and optimization methods to select the best parameters for improved cutting tool performance. However, in most of these cases the process used during machining was turning.

It is of considerable economic importance for production engineers to know in advance the machinability of a work material to maintain efficient and economical manufacturing. The wide spread application of MMCs will not be possible without finding a solution for the shortened tool life encountered during cutting operations. Even with near-net shape manufacturing methods (i.e., work piece is close to final form), the need for machining cannot be completely eliminated.

4.2.1 Chip Formation

Understanding chip formation is very important to comprehend the mechanics of cutting and the cutting forces. During turning of conventional alloys, the surface of a circular piece is removed by bringing the cutting edge of a tool against it while the piece is rotating. The material ahead of the cutting tool edge is subjected to severe plastic deformation and subsequent shearing results in chip formation. Chip formation during turning of metal matrix composites differs slightly in some aspects. The reinforcement particles or fibers are distributed randomly at and about the tool movement/edge. The presence of hard reinforcements alters the plastic deformation characteristics of the soft matrix material compared to those of a conventional alloy. Thus, the change in mechanical properties coupled with reinforcement, configuration and distribution in the matrix determines the mechanism of chip formation (shearing, plowing, particle interface debonding, pull out and cracking) and hence the machinability of MMCs.

4.2.2 Wear Mechanisms and Tool Life

For a given tool and cutting condition, the tool wear rate will depend on the mechanical properties of composites. Hence, it is clear that the reinforcement morphology, distribution and volume fraction, as well as the matrix properties, are all factors that affect the overall cutting process. Since abrasion is the primary source of tool wear at different feed rates, the recommendations are to use high-feed rates and depths of cut during roughing operations. Several researchers have also indicated that polycrystalline diamond (PCD) tools are the only tool material that is capable of providing a useful tool life during the machining of particulate light metal MMCs. PCD is sufficiently harder than most of the ceramic reinforcements and has no chemical tendency to react with the workpiece material. Furthermore, PCD tools contain larger grain structures that withstand more abrasion wear by micro-cutting compared to tools with a smaller grain structure. A comprehensive collection of such information available in the public domain is summarized in the following paragraphs.

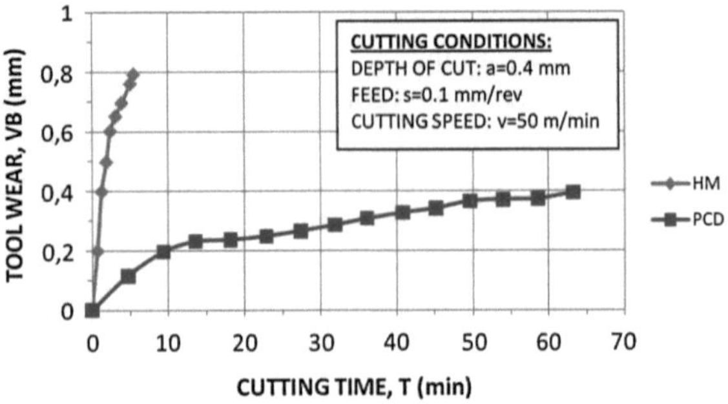

Fig. 4.1 Wear progression curve for multiple coated tool carbide tool vs. PCD tool [6]

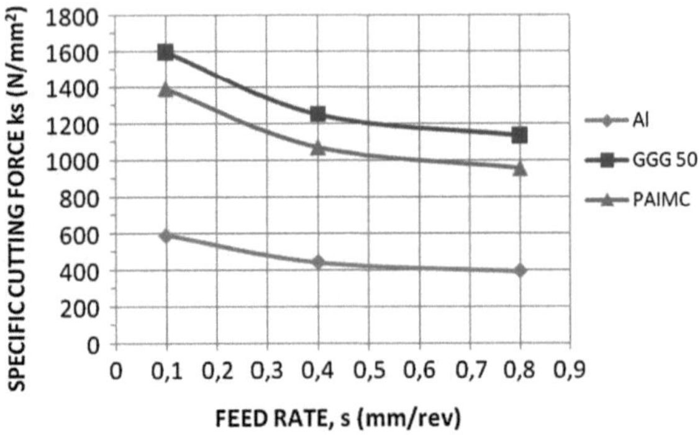

Fig. 4.2 Specific cutting force with feed rate [6]

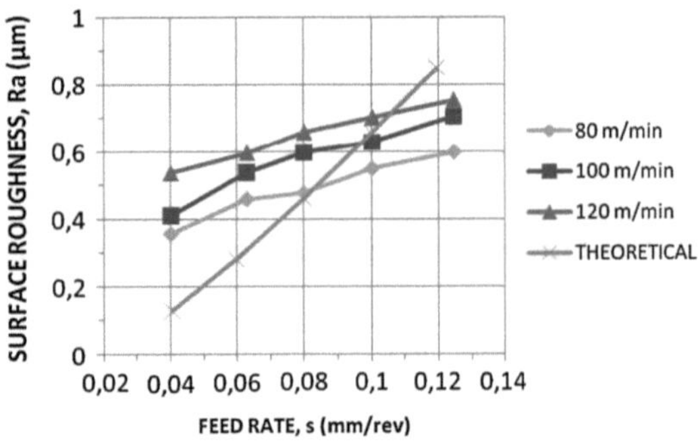

Fig. 4.3 Surface roughness vs. feed rate with different cutting speed [6]

Tomac and Tonnessen [6] investigated the machinability of Al-SiC MMCs using PCD, chemical vapor deposition (CVD), and coated tungsten carbide tools. The investigation revealed that abrasive wear is the main mode of tool failure. The PCD tools had over 30 times higher tool life than carbides they used under similar cutting conditions, as shown in Fig. 4.1. In addition, the CVD tools were better than other less hard cutting tools except PCD. As shown in Fig. 4.2, tool wear and cutting force decreased with increase in feed rate due to thermal softening of the MMCs. In addition, the machined surface finish deteriorated with increasing cutting speed, as shown in Fig. 4.3. The researchers have noted that during rough machining, stable built-up edge (BUE) formation on the cutting tool edge was

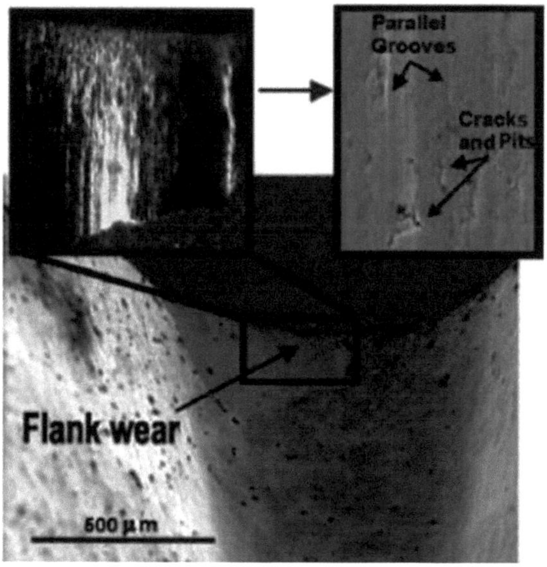

Fig. 4.4 Typical modes of tool wear [8]

observed. Similar to the work of Tomac and Tonnessen [6], Hung et al. [7] carried out tests on turning of MMCs with CBN, PCD, WC and DCC (diamond-coated carbide) tools and compared the tool wear with different turning parameters.

More recently Kishawy et al. [8] presented an analytical model for predicting tool flank wear progression during turning of particulate reinforced MMCs. Figure 4.4 shows a typical mode of tool wear when cutting MMC. A methodology was proposed for analytically predicting the wear progression as a function of tool/workpiece properties and cutting parameters. The following equation was developed to determine the flank wear:

$$\frac{dV_B}{dt} = \frac{3K}{4RSin\left(\frac{\psi}{2}\right)V_B \tan\alpha}\left[\sqrt{\left(\frac{N\vartheta_c}{H_t}\right)^2 + \left(\frac{N\vartheta_c}{xH_t}\left(\frac{H_t}{H_a}\right)^k\left(\frac{D}{d}\right)\left(\frac{f_v}{z(1-f_v)}\right)\right)^2}\right] \tag{4.1}$$

According to their model the wear mechanisms that were identified during cutting MMCs were two body and three body abrasions. Their proposed model generally agrees with the measured tool wear data depicted in Fig. 4.5.

In a search for a better tool performance Chen [9] studied the performance of self-propelled rotary tools through turning trials on Al-SiC MMCs and compared the results with fixed tools. It was concluded that the rotary tool showed better tool life and wear resistance. Coelho et al. [10] proposed an empirical relationship between tool wear and milling speed and agreed with Chen's findings that the cutting speed had negligible effect on tool wear.

On the contrary, several researchers have reported that the tool life decreases with increase in cutting speed. Lin et al. [11] reported a similar phenomenon as

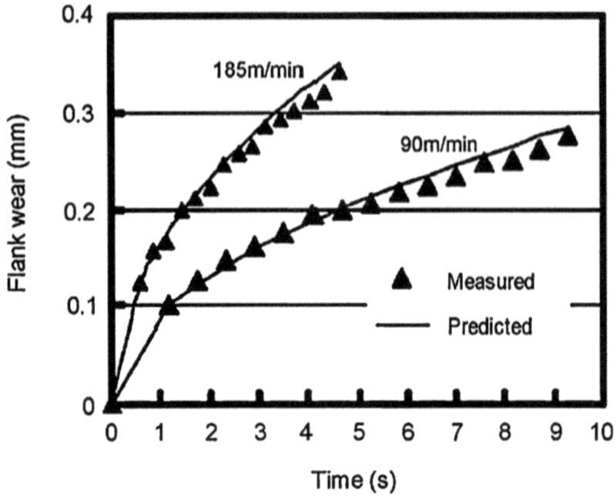

Fig. 4.5 Wear progression during cutting A356 MMC at different cutting speeds ($f = 0.3$ mm, depth of cut 0.1 mm, Al alloy 20% SiC) [8]

Sahin [12] when machining Al alloy with 20% reinforcement using PCD tools where the wear increased rapidly with the increase in cutting speed. They studied the machinability of MMCs and the dependence of tool wear on the volume fraction of reinforcements in the MMCs. It was observed that the abrasive tool wear accelerated when the volume fraction of the reinforcements in the MMC exceeded a critical value. Although the tool wear was increased, the tests showed that surface finish remained constant with increase in speeds (at high speeds 300–700 m/min).

In 1997, Tonshoff and Winkler [13] conducted a study on magnesium-based MMC with 20% SiC reinforcement particles using *TiN*-coated, PCD coated and PCD-tipped tools. They reported that the *TiN* coatings have shorter life and the PCD-coated tools showed good performance before the deterioration of the coating film. Compared to coated PCD tools, PCD-tipped tools showed better results. El-Gallab and Sklad [14, 15], studied the performance of PCD tools during turning MMCs. Grooves on the tool face along the chip flow direction were observed. The grooves on the rake face filled with smeared work material and formed a built-up edge, which seemed to be beneficial since it protected the tool rake from further abrasion, as shown in Fig. 4.6. However, for all the tested tools, the tool life was limited by excessive flank wear due to abrasion and micro-cutting of tool material.

Weinert [16] studied tool wear progression during turning of MMC with carbide tool and PCD tool. The tested materials were 10 and 20% volume fraction aluminum-based MMC reinforced with Al_2O_3, SiC, and B_4C. It was found that the reinforcement hardness was a dominant factor for tool wear. The study showed the higher tool wear when cutting 10% B_4C MMCs compared to Al_2O_3 short-fiber reinforced Al alloy (Fig. 4.7). They also observed better performance of PCD tools compared to carbide tools.

Figure 4.8 shows the progression of flank wear of PCD tools during turning of different aluminum MMCs. Yanming and Zehua [17] also investigated the tool wear during machining *SiC* particle reinforced aluminum matrix composites.

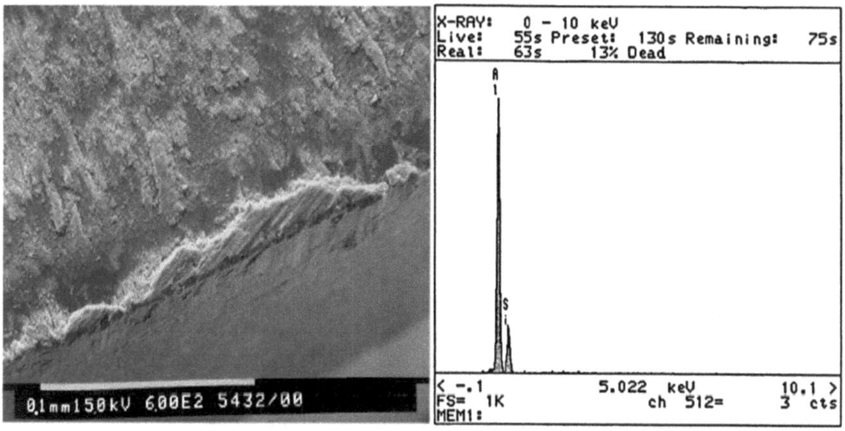

Fig. 4.6 Built-up edge on PCD tools. *Left*: $v = 670$ m/min, $f = 0.25$ mm/rev, depth of cut = 2.5 mm; *right*: X-ray dispersion of built-up edge [14]

Fig. 4.7 Tool wear of carbide tools when turning Al alloy matrix MMC materials at 100 m/min cutting speed [16]

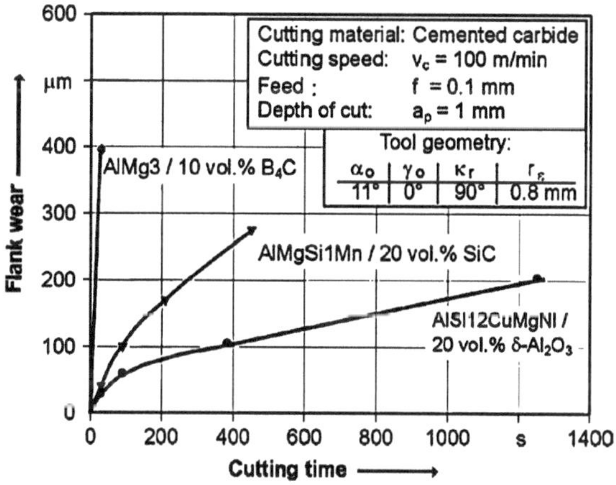

The main objective of this work was to study the effect of material properties on tool wear mechanisms. Volume fraction and the size of *SiC* particles played an important role on tool life. They concluded that coarser reinforcements and higher volume fractions largely influenced the tool performance and required cutting tools with high hardness.

Barnes et al. [18] studied the hot machining of MMCs. The material was preheated to 200–400°C. At low-cutting speeds the tool life was found to be improved when MMC was preheated and the reason was attributed to the formation of built-up edges (BUEs). Ding et al. [19] studied the machinability characteristics with various polycrystalline boron nitride (PCBN) and PCD tools. With 0.3 mm depth of cut and feed rate of 0.1 mm/rev, the experiments were carried out at the cutting speed of 50 and 400 m/min. Compared to PCBN tools,

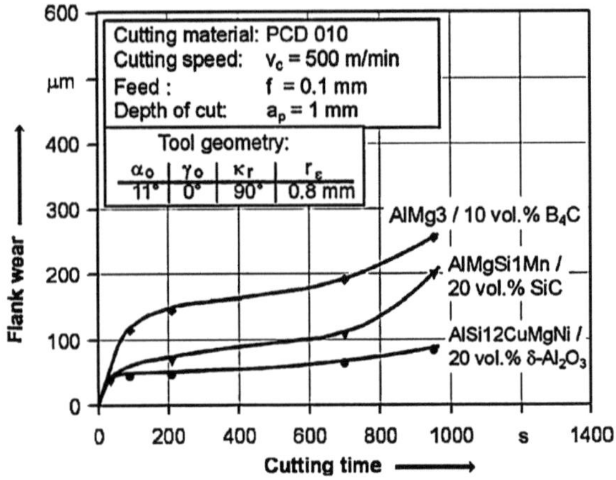

Fig. 4.8 Tool wear of PCD tools when turning Al alloy matrix MMC materials at 500 m/min cutting speed [16]

the improved tool life was found when using PCD tools. The poor performance of PCBN tools was attributed to the observed groove wear.

Diamond is mainly used in two forms; primarily brazed PCD or CVD. CVD diamond is a newly developed super-hard tool material made up with a pure diamond coating over a carbide substrate, and has become a competitive alternative to PCD tools for two reasons; (1) although PCD tools demonstrate superior tool life among other tools, they are commonly regarded as economically non-sustainable due to their high cost, and (2) the hardness of CVD (10,000HV) diamond is much higher than that of PCD (6000HV). Using one millimeter thick layer of CVD cutting tools, Weinert et al. [20] carried out turning experiments on brake drums made from aluminum alloy reinforced with 20% SiC. Compared to the traditional PCD tools, CVD tools showed improved tool performance in terms of tool life.

4.2.3 Studies on Surface Roughness

Many researchers have studied the relation between surface roughness and the cutting speed. Some of the obtained results were complementary while some showed a contradiction to the remaining available data. Manna and Bhattacharya [21], carried out turning of Al alloy with 15% *SiC* MMC using tungsten carbide tools. It was found that the surface roughness decreased by half with the increase in the cutting speed from 60 to 180 m/min.

It is commonly believed that the feed rate has a negative effect on surface integrity, as shown in Fig. 4.9. The average roughness can be estimated with the following theoretical equation:

$$R_a = \frac{f^2}{18\sqrt{3} \times R} \tag{4.2}$$

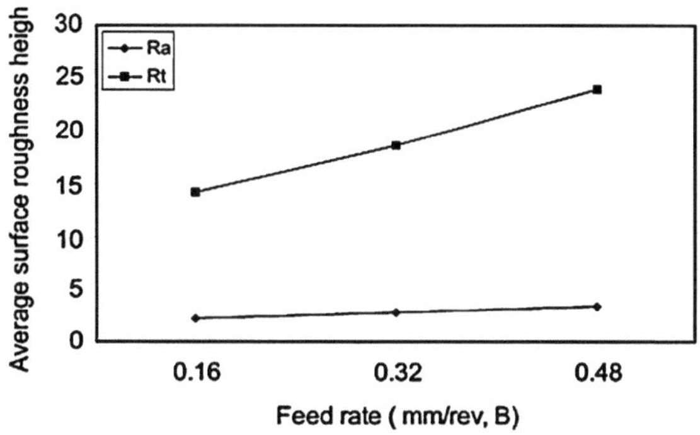

Fig. 4.9 Average surface roughness vs. feed rate while machining different MMCs [21]

As noted by Tomac and Tonnessen [6], when feed rate is low, the theoretically calculated surface roughness (R_a) overestimates the real value measured from the experiments. This low roughness was attributed to the large nose radius when the tool wears. Similar results are reported by Manna and Bhattacharya [21] while machining *Al-SiC* MMC.

In a different study, Ciftci et al. [22] investigated the effect of particle size on surface roughness when turning *Al/SiCp* MMC with coated and uncoated tools. Based on the conducted study they concluded that roughness values (R_a and R_t) increased with the increase in particle size and the volume fraction of the reinforcement particles. Also, they observed better improvement of the surface finish at lower cutting speed when using un-coated carbide tool compared to cutting with coated tools (*TiC*, Al_2O_3). At higher cutting speeds a reversed trend was reported where the coated tools outperformed the uncoated ones.

Songmene and Balazinski [23] studied the machinability of a new family of MMCs "*GrA-Ni*" consisting of an aluminum matrix reinforced with nickel-coated graphite particles and *SiC* or Al_2O_3 particles. This generation of MMC is usually used for cylinder sleeves and brake drums. The researchers conducted turning tests using PCD and diamond-coated carbide tools. The results indicated that the surface roughness was lower when using PCD tools.

They also observed that the cutting speed did not influence the average roughness, R_a, significantly and their observations were attributed to the absence of a BUE in the range of cutting speed. Therefore, it was concluded in their study that the surface integrity is independent of cutting speed at higher cutting speed regime in which the BUE is absent.

4.2.4 Ultrasonic Vibration Turning

Liu et al. [24] conducted ultrasonic vibration turning tests on Al/SiC MMC. By comparing with the conventional turning, their investigation concluded that lower cutting force was generated when using ultrasonic vibration process but the effect

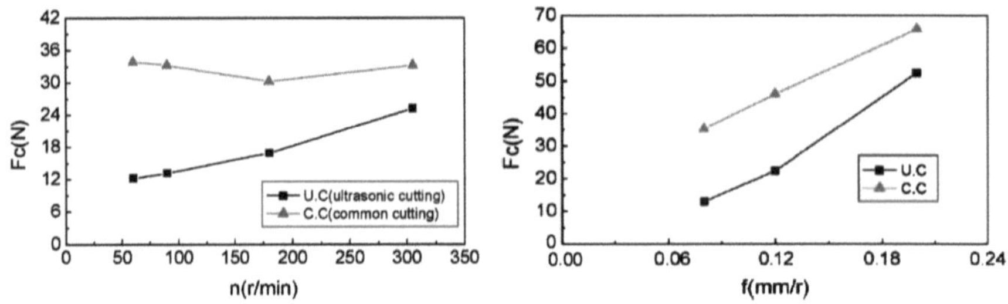

Fig. 4.10 Cutting force vs. cutting speed and feed rate [24]

was more pronounced at lower values of cutting speed and feed rate, as shown in Fig. 4.10. The difference in the characteristics of the generated forces is attributed to the absence of BUE when ultrasonic turning is employed.

4.2.5 Modeling of the Process

Most of the early studies were based either on experimental studies that compare the performance of different tools or on the empirical and numerical studies related to tool life. Very little research has been conducted to predict the cutting forces generated in MMC machining. Hoecheng et al. [25] carried out a thorough study on the effect of cutting conditions, such as cutting speed, depth of cut, rake angle and cutting fluid during machining MMCs. The volume fraction of reinforcing particles was responsible for the increased cutting force, while decreasing negative rake angle can lead to a reduction in cutting force.

First analytical force model was developed by kishawy et al. [26] where the cutting force was estimated based on the energy consumed in the primary, secondary shear zone and reinforcement particle displacement and fracture. According to the model, total energy per unit volume of metal removed is:

$$e = E_p + E_s + E_d \tag{4.3}$$

The energy consumed in the secondary deformation zone was assumed as one-third of that in the primary shear zone. Although this assumption is true for alloys, it is questionable for the case of MMCs. It should be mentioned here that energy due to plowing was not considered.

Pramanik et al. [27] developed an analytical model where plowing force was considered. The chip formation force was obtained by using Merchant analysis but those due to matrix plowing deformation and particle fracture were formulated respectively with the aid of the slip-line field theory of plasticity and the Griffith theory of fracture. However, the chip–tool friction force due to reinforcement particles was not considered. Recently, Davim et al. [28] extended the classical Merchant's theory of metal cutting to machining of MMCs. They concluded that while machining MMCs, Merchant's prediction of shear angle was an overestimate of the observed shear angle. More recently, Dabade et al. [29] presented an analytical model by considering the friction at the chip–tool interface. However, the authors did not consider the effect of particle debonding and plowing force.

4.2.6 Numerical Modeling

Many attempts have been made to model the process of turning MMCs by using finite-element analysis (FEA). Monaghan and Brazil [30] modeled the machining (using FORGE2 code) of A356 aluminum alloy. Different regions of the chip and the machined surface were then submodeled (using ANSYS software), to study the non-uniform matrix flow, tool wear, failure of particle–matrix interface, change of loading (in different sub-models) and the generated residual stress in MMC surface due to the application of pressure and temperature obtained from machining simulation of aluminum alloys.

El-Gallab and Sklad [31] followed similar modeling procedure as that in the work by Monaghan and Brazil [30] except that force and temperature data applied on the MMC surface were obtained from MMC machining tests. By evaluating the tool wear and sub-surface damage parameters from a set of experimental results, the main objective of this work was to select the optimal cutting parameters that would lead to the best surface quality and the longest tool life. The surface quality was examined based on the ratio of the hydrostatic stress beneath the machined surface to the shear flow strength obtained from finite-element simulation.

Ramesh et al. [32] carried out a transient dynamic finite-element analysis on the diamond turning of an Al6061/SiC MMC. The range of forces and stresses that could be generated during micro-machining of MMC elastic–plastic was presented.

Zhu and Kishawy [33] developed a plane-strain thermo- finite-element model to simulate orthogonal machining of $Al6061/Al_2O_3$ composite using a tungsten carbide tool. They studied temperature distribution in the matrix macroscopically (Fig. 4.11), and reported the average values of shear stresses (Fig. 4.12) in the matrix and on particles at different locations in the chip and primary/secondary deformation zones.

Pramanik et al. [27] developed the stress and strain fields in an MMC and analyzed the tool wear, particle debonding, displacements and inhomogeneous deformation of the matrix material. It was found that the particle debonding, surface damage and the tool wear were fundamentally affected by the mode of stress/strain distributions in the particle/matrix and the tool–particle interactions.

Dandekar and Shin [34] developed a 3D finite-element model to study the subsurface damage of *A359/SiC/20p* MMC, by releasing the assumption that the particles are perfectly bonded to the matrix. In their study, the modeling of MMC machining behavior was realized by a two step approach. In the first step, the MMC was considered as an equivalent homogeneous material (EHM). The cutting

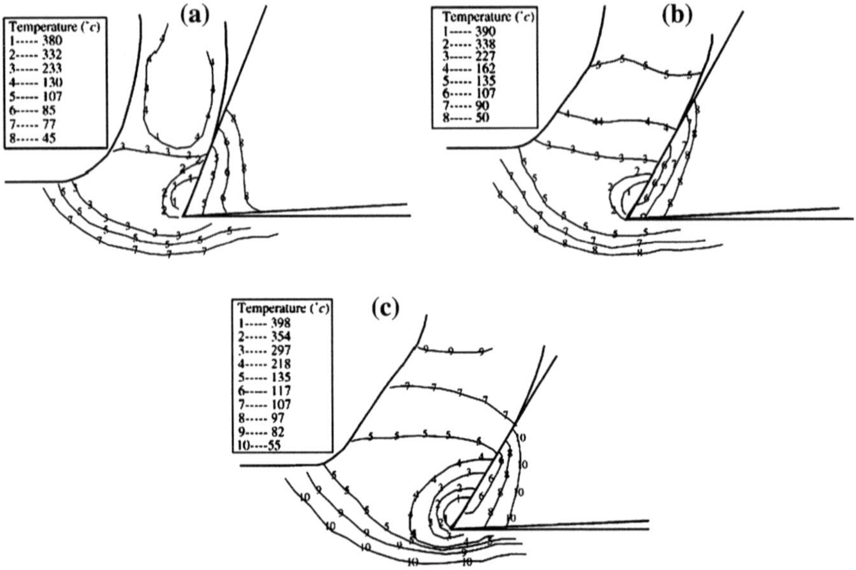

Fig. 4.11 Temperature distribution on the matrix. **a** Temperature contour for feed = 0.1 mm/rev. **b** Temperature contour for feed = 0.2 mm/rev. **c** Temperature contour for feed = 0.3 mm/rev [33]

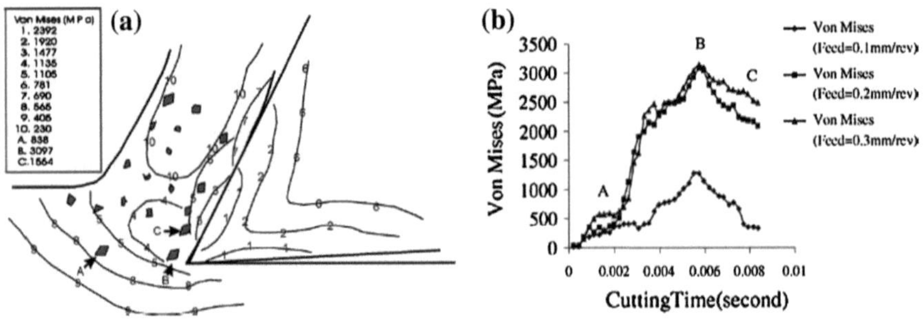

Fig. 4.12 Von Mises stress distribution on the matrix and hard particle. **a** Contours of Von Mises equivalent stress distribution for feed = 0.2 mm/rev. **b** Von Mises stress distribution on the particles [33]

forces, temperatures and stress distributions were obtained in a macroscopic manner. In the second step, the obtained temperature and stress distributions were utilized as the inputs to a micromechanics approach, in which the properties of matrix, particulate and particulate–matrix interface were prescribed individually. The matrix was considered as isotropic thermo-elastic–plastic material and the particulate was assumed to be linear elastic before fracture. The particulate–matrix interface was represented by a layer of cohesive zone. In such a manner, the deformation behavior of matrix and particulate can be described by the continuum elements, while the debonding and subsequent sub-surface damage can be predicted by the behavior of the cohesive zone elements. They compared the results obtained from the EHM machining model with the experimental data in terms of the measured cutting force and thrust force; good agreement was obtained. It should be noted here that in using the EHM model, the tool–particle interaction was neglected. However, the good agreement may suggest that considering MMC as a equivalent homogeneous material works well in terms of predicting macroscopic processing parameters such as cutting forces, temperatures, stress distributions, etc. To investigate the sub-surface damage, three regions on the machined surface below the cutting edge were selected. The actual sub-surface damage was measured with SEM images as the depth of damage. The results indicated that the extent of damage is not sensitive to the cutting speed and can basically be expressed as a function of the feed rate.

4.3 Drilling of MMC

Considerable research in the field of drilling MMC had been conducted to improve tool life, and optimize the cutting conditions for different drills. Commonly used drills in these studies include PCD tipped drills, high-speed steel (HSS) drills, diamond-coated HSS drills, coated carbide drills (tungsten carbide (*WC*) and *TiN*), and *TiAlN* coated drills.

Cronjager and Meister [35, 36] studied the effect of cutting speed on the tool wear of two different drills. It was found that the tool wear of *WC* drills increases with the increase in the drilling speed. However, the drilling speed has no evident effect on the wear of PCD twist drills in the range of low-to-medium cutting speeds of 15–300 m/min. It was reported by Coelho et al. [37] that the carbide drills are not proper tools for aluminum alloy 2,628 with 15% *SiC* reinforcements. Again in 1994, Coelho et al. [38], presented results of machinability tests on Al-MMC with 15% *SiC* reinforcements using PCD drill bits. The experimental work mainly concentrated on conventional speed drilling and high-speed drilling. Coelho et al. [38] continued their attempts by developing PCD tipped drill bits to drill MMC and presented a comparison among the results of different drill tests including the PCD drill bits and other tools such as HSS, diamond-coated HSS, *WC* and *TiN*-coated carbide tools. The tests were carried out on Aluminum alloy 2,628 with 15% *SiC* reinforcements. The results in Fig. 4.13 show that PCD-tipped drill bits perform the best under different cutting conditions among all the drill materials. They also reported that the cutting speed was not a significant factor affecting tool life which confirms the earlier findings reported by Lane [39].

Jawaid et al. [40] carried out drilling tests on aluminum alloy 2,014 with *SiC* reinforcements. It was found that thrust force decreased, however, the tool wear increased with the increase in the cutting speed.

Venkatesh et al. [41] performed drilling tests on the aluminum alloy reinforced with *SiC* particles (volume fractions 12 and 20%), with a new type of micro-drill

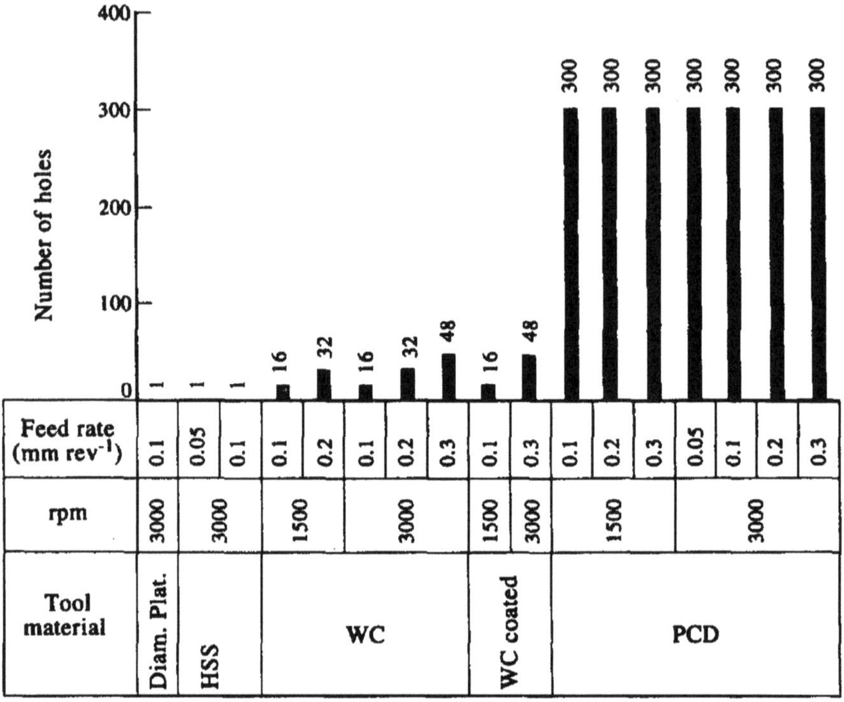

Fig. 4.13 Results of drilling experiments in machining MMC [39]

made of fine carbide grains (2 μm). It was found that the holes drilled with the lower feed rate had better peripheral waviness for both types of tested MMCs. Results also confirmed what was already pointed out for the case of turning of MMC where higher particle volume fractions increase tool wear.

Narutaki [42] investigated the effect of cutting conditions on tool wear during drilling 6061 *Al* MMC reinforced with *SiC* and Al_2O_3. It was discovered that flank wear decreased with an increase in feed rate due to the decrease of total cutting length. The wear was observed to be almost independent of spindle speed.

Songmene and Balazinski [43], conducted drilling tests using HSS twist drill to study the machinability of *GrA-Ni*® MMC. Better machinability of *GrA-Ni*® was reported than all other *SiC* particle reinforced MMCs. According to Songmene et al. [44] *GrA-Ni*® has better machinability because ceramic and graphite particles found in *GrA-Ni*® are not as hard as *SiC* or Al_2O_3 particles. In addition, graphite flakes present in the aluminum matrix act as a solid lubricant during drilling. From the observation of the formed chips, the authors confirmed the function of reinforcing particles as the 'chip breaker'. The continuous chip form of the *GrA-Ni 6 s-2.5G*, similar to that of *Al380*, was attributed to the lower total volume fraction of particles and the consequent smaller possibility to break the chip. However, from the highly serrated chip of *GrA-Ni 5A.4G*, in which the total volume of particles is quite similar to that in *GrA-Ni 6 s-2.5G*, the dominant mechanism

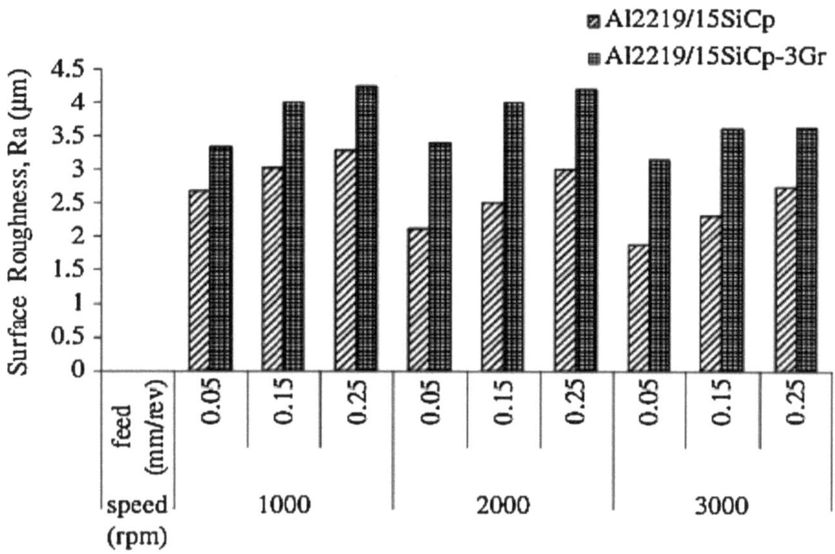

Fig. 4.14 Variation of surface roughness for both the materials Al2219/15SiCp and Al2219/15SiCp-3Gr when drilled using carbide drill [45]

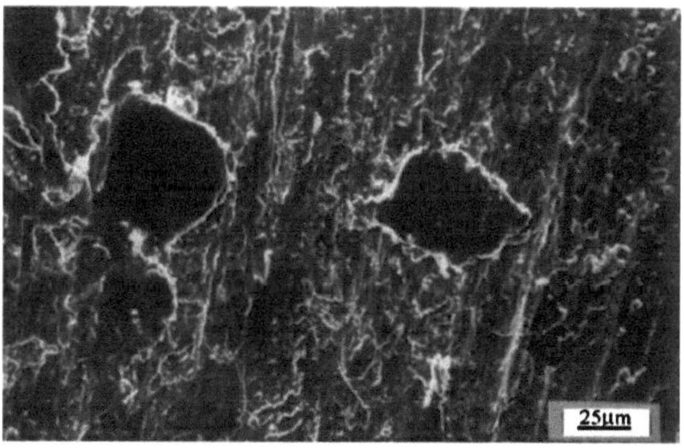

Fig. 4.15 SEM image of the particle pullout on the surface drilled with carbide drill on Al2219/15SiCp-3Gr, $V = 3{,}000$ rpm and feed of 0.15 mm/rev [45]

accounting for the chip form during machining MMCs still needs further investigations.

Tosun and Muratoglu [45] investigated the surface integrity and analyzed the subsurface when drilling *Al 2,124/17 SiCp* MMCs. It was concluded that the surface quality can be improved by increasing the drill hardness.

Basavarajappa et al. [46] conducted an experimental study on surface roughness variations on the drilled surfaces of *Al2219/15SiCp* and *Al2219/15SiCp-3Gr* MMC using carbide and coated carbide drills. The surface roughness was found to be decreased with the increase in cutting speed and increased with the increase in feed rate (Fig. 4.14). Their SEM observations on the drilled surface showed the existence of micro-cracks, particle pull out and shearing of particles (Figs. 4.15, 4.16).

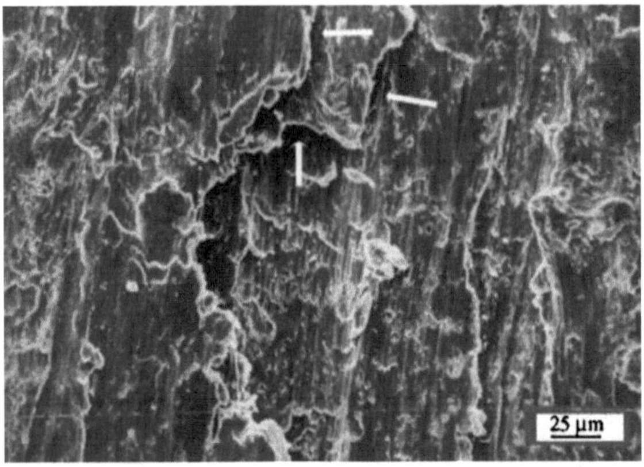

Fig. 4.16 Micrograph shows the existence of micro-cracks (indicated by *arrows*) on the surface of the drilled hole on Al2219/15SiCp-3Gr with a cutting speed of 3,000 rpm and feed rate 0.15 mm/rev using carbide drill [45]

4.4 Milling of MMC

One of the earliest studies in milling of MMC was carried out by Lane [39] on newly casted MMC with different types of tools; namely HSS and coated tools. These tools were proved to be not economically justifiable for the task, however, *TiC* or *TiN* coatings offered slight advantage. Following these trials and in an attempt to find better tool performance, Lane and Finn [47] studied the performance of different CVD tools with thin and thick films. It was reported that CVD tools with thin films failed catastrophically during the end milling of *Al*-based MMC with 20% *SiC*. This tool failure was attributed to coating spilling and the consequent damage to the relatively soft carbide substrate.

Songmene and Balazinski [43] conducted dry milling tests using *TiCN*-coated carbide inserts under various cutting conditions to study the machinability of *Al*-based MMC materials. Different types of reinforcement were utilized in this investigation including *GrA-Ni 5A.4G* (5 vol% Al_2O_3 4 vol% *Ni-Gr*), *GrA-Ni 10S.4G* (10 vol% *SiC*, and 4 vol% *Ni-Gr*), *GrA-Ni 6 s-2.5G* (6 vol% *SiC* and 2.5 vol% *Ni-Gr*), *Al–Si.20SiC* (20 vol% *SiC*) and pure aluminum *Al* 380. The objective of this study was to evaluate the milling tool performance in terms of the progression of flank wear and the tool life and to provide the database of the effects of hardness, size and volume fraction of the reinforcing particles on the MMCs' machinability.

In Fig. 4.17, the progressions of the tool flank wear which generated when milling different materials at a cutting speed of 61 m/min. It can be noted that the higher the volume fraction of reinforcing particles, the faster the tool wear progresses. This is clear evidence given the non-ignorable role that the tool–particle interaction plays during machining MMCs. It can also be seen that the rates of tool wear for all the composites containing *Ni*-coated graphite particles are lower than that of the composite with *SiC* particles only. During the milling of composites containing *SiC* particles, with hardness higher than *TiCN* tool coatings, the tool–workpiece interaction is dominated by grinding and therefore *Al-9Si.20SiC* shows the poorest machinability in terms of tool wear. However, the improved

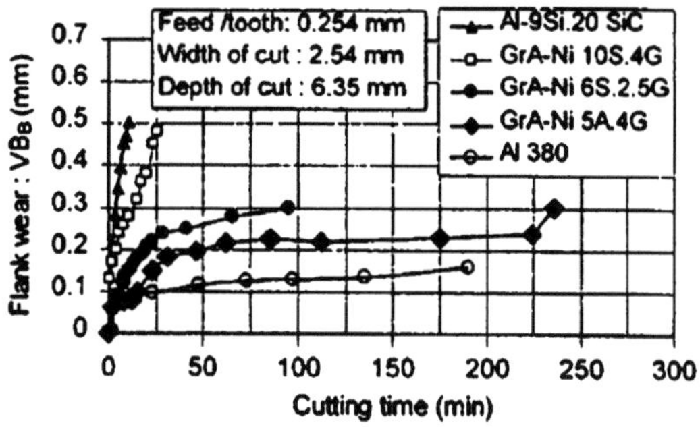

Fig. 4.17 Wear progression at 61 m/min cutting speed [43]

machinability by introducing nickel-coated graphite particles was attributed to the lubricating nature of the graphite particles. Among those containing graphite particles is *GrA-Ni 5A.4G* which demonstrated the best machinability due to the low hardness of alumina particles.

Figure 4.18 shows the tool life-cutting speed relations for the studied composites under the framework of Taylor's tool life model. Compared with conventional alloys, the higher Taylor exponents ($0.69 < n < 0.95$) of these curves indicate the less impact of the cutting speed on the tool life. In fact, the greater effect of the feed rate than the cutting speed on the tool life during machining graphitic *SiC* reinforced MMCs is additional evidence supporting the intense interaction between particles and the cutting tool. The abrasive load at the tool–workpiece interface is proportional to the chip load. Increasing the feed rate inevitably increases the abrasive load and aggravates the grinding effect of the hard particles on the tool flank face. To conclude the findings of this study, the total volume fraction and the hardness of reinforcing particles are clearly shown to be the major factors influencing the machinability of MMCs. Decreasing the total density of the reinforcement and replacing the harder particles (*SiC*) with softer ones (Al_2O_3) has largely improved the machinability. However, since the total particle volume fraction of each composite in the comparison (Fig. 4.17) is not the same, the claim that the improved machinability of *GrA-Ni 5A.4G* is caused by the incorporation of graphite or the reduction of total particle volume fraction and the introduction of softer particles is not clear. Further studies are needed to quantify the lubricating effect of the graphite particles on machinability.

Chandrasekaran and Johansson [48], performed dry face milling tests on aluminum-based MMCs reinforced with *SiC* particles. The experiment was carried out using carbide tools and PCD tools. With constant depth of cut (1.5 mm), two different feed rates, low- and high-cutting speeds (15 and 1, 178 m/min) were chosen to evaluate the tool wear and surface quality for MMCs with different *SiC* particle fractions. The results indicated that the cutting speed has a minor effect on the tool flank wear rate and the tool life. However, feed rate, unlike conventional materials, shows a noticeable effect. Furthermore, the flank wear was most severe in both cases for the material with 38% *SiC* reinforcement, suggesting that the higher the volume fraction of reinforcing particles, the shorter the tool life.

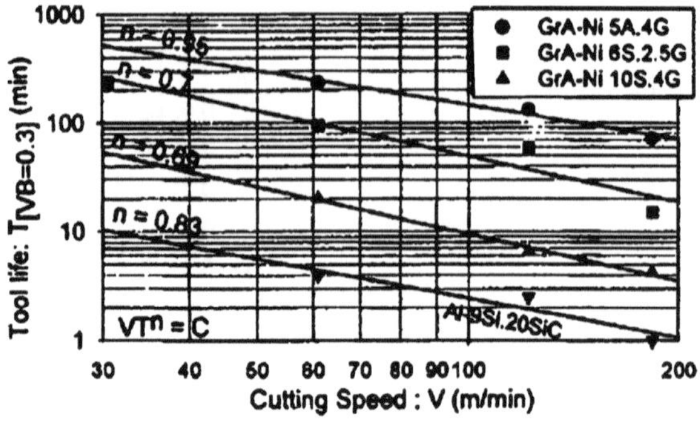

Fig. 4.18 Tool life as function of cutting speed [43]

4.5 Conclusion

Several machinability studies have been conducted on different types of MMC materials. The objective of these studies is to define the boundaries for safe and economical cutting conditions which can be used to machine MMCs. In all the aforementioned studies, the poor machinability index was attributed to the abrasive nature of the reinforcements and it has been commonly believed that the properties (especially hardness), the size, and the volume fraction of the reinforcing particles, are the main factors that influence the machinability of all kinds of MMCs. It has been qualitatively justified that in general the machinability decreases with the increase in the hardness and the volume fraction of the particles. Analytical and numerical studies have been carried out in an effort to quantify these effects and some satisfactory results were obtained. However, from a mechanics point of view, as for the conventional alloys (homogeneous matrix material only) the deformation behavior and the friction law governing chip–tool interface friction during machining has not yet been well defined. In addition, the effect of reinforcement particles on the deformation mechanisms during machining of MMCs is also not fully understood. There is still research required to clarify these interactions before we can shed light on new possible concepts and further developments regarding MMCs and machining thereof.

References

1. Rohatgi PK, Asthana R, Das S (1986) Solidification, structures, and properties of cast metal-ceramic particle composites. Int Mater Rev 31:115–139
2. Hashim J, Looney L, Hashmi MSJ (2002) Particle distribution in cast metal matrix composites—Part I. J Mater Process Tech 123(2):251
3. Zhou W, Xu ZM (1997) Casting of SiC reinforced metal matrix composites. J Mater Proc Tech 63(1–3):358
4. Suresh S, Mortensen A, Needleman A (1993) Fundamentals of metal–matrix composites. Butterworth-Heinemann, UK

5. Chawla KK, Chawla N (2006) Metal-matrix composites. Kirk-Othmer encyclopedia of chemical technology. Wiley, NJ
6. Tomac N, Tannessen K, Rasch FO (1992) Machinability of particulate aluminium matrix composites. CIRP Annals Manuf Tech 41(1):55–58
7. Hung NP, Venkatesh VC, Loh NL (1998) Cutting tools for metal matrix composites. Key Eng Mater 138–140:289–326
8. Kishawy HA, Kannan S, Balazinski M (2005) Analytical modeling of tool wear progression during turning particulate reinforced metal matrix composites. CIRP Annals Manuf Tech 54(1):55–58
9. Chen P, Hoshi T (1992) High-performance machining of sic whisker-reinforced aluminium composite by self-propelled rotary tools. CIRP Annals—Manuf Tech 41(1):59
10. Coelho RT et al. (1993) Conventional machining of an aluminium based Sic reinforced metal matrix composite (MMC) alloy. In: Proceedings of 30th Matador, Manchester
11. Lin JT, Bhattacharyya D, Lane C (1995) Machinability of a silicon carbide reinforced aluminium metal matrix composite. Wear 181–183(Part-2):883
12. Sahin Y (2004) Preparation and some properties of SiC particle reinforced aluminium alloy composites. Mater Des 24(8):671
13. Tonshoff HK, Winkler J (1997) The influence of tool coatings in machining of magnesium. Surf Coat Tech 94–95:610
14. El-Gallab M, Sklad M (1998) Machining of Al/SiC particulate metal-matrix composites. Part I: Tool performance. J Mater Process Tech 83(1–3):151–158
15. El-Gallab M, Sklad M (2000) Machining of Al/SiC particulate metal matrix composites. Part III: Comprehensive tool wear models. J Mater Process Tech 101(1–3):10
16. Weinert K, Konig W (1993) A consideration of tool wear mechanism when Machining Metal Matrix Composites (MMC). CIRP Annals—Manuf Tech 42(1):95–98
17. Yanming Q, Zehua Z (2000) Tool wear and its mechanism for cutting SiC particle-reinforced aluminium matrix composites. J Mater Proc Tech 100(1–3):194
18. Barnes S, Pashby IR, Mok DK (1996) The effect of workpiece temperature on the machinability of an aluminum/SiC MMC. J Manuf Sci Eng 118(3):422–427
19. Ding X, Liew WYH, Liu D (2005) Evaluation of machining performance of MMC with PCBN and PCD tools. Wear 259(7–12):1225
20. Weinert K et al (2001) Spanende bearbeitung von bauteilen aus al matrix verbundwerkstoffen. Materialwissenschaft und werkstofftechnik 32(5):447–461
21. Manna A, Bhattacharayya B (2003) A study on machinability of Al/SiC-MMC. J Mater Process Tech 140(1–3):711
22. Ciftci I, Turker M, Seker U (2004) Evaluation of tool wear when machining SiCp-reinforced Al-2014 alloy matrix composites. Mater Des 25(3):251–255
23. Songmene V, Balazinski M (2001) Machining of a Graphitic Sic-Reinforced Aluminium Metal Matrix Composites with Diamond Tools. CIRP International Seminar on Program in Innovative Manufacturing Engineering
24. Liu CS, Zhao B, Gao GF (2002) Research on the characteristics of the cutting force in the vibration cutting of a particle-reinforced metal matrix composites SiCp/Al. J Mater Process Technol 129(1–3):96–199
25. Hocheng H et al (1997) Fundamental turning characteristics of a tribology-favored graphite/ aluminum alloy composite material. Comp A Appl Sci Manuf 28(9–10):883
26. Kishawy HA, Kannan S, Balazinski M (2004) An energy based analytical force model for orthogonal cutting of metal matrix composites. CIRP Annals Manuf Tech 53(1):91–94
27. Pramanik A, Zhang LC, Arsecularatne JA (2006) Prediction of cutting forces in machining of metal matrix composites. Int J Mach Tool Manuf 46(14):1795–1803
28. Davim JP, Silva J, Baptista AM (2007) Experimental cutting model of metal matrix composites (MMCs). J Mater Process Tech 183(2–3):358
29. Dabade UA, Dapkekar D, Joshi SS (2009) Modeling of chip–tool interface friction to predict cutting forces in machining of Al/SiCp composites. Int J Mach Tool Manuf 49(9):690–700

30. Monaghan J, Brazil D (1998) Modelling the flow processes of a particle reinforced metal matrix composite during machining. Composites Part A: Appl Sci Manuf 29(1–2):87
31. El-Gallab M, Sklad M (2004) Machining of aluminum/silicon carbide particulate metal matrix composites: Part IV. Residual stresses in the machined workpiece. J Mater Process Tech 152(1):23
32. Ramesh MV et al (2001) Finite-element analysis of diamond turning of aluminium matrix composites. Comp Sci Technol 61(10):1449
33. Zhu Y, Kishawy HA (2005) Influence of alumina particles on the mechanics of machining metal matrix composites. Int J Mach Tool Manuf 45(4–5):389
34. Dandekar CR, Shin YR (2009) Multi-step 3-D finite element modeling of subsurface damage in machining particulate reinforced metal matrix composites. 15th Fr Natl Conf Compos 40(8):1231–1239
35. Cronjager L, Meister D (1991) Machining of fibre and particle-reinforced aluminium. CIRP Annals—Manuf Tech 41(1):63–66
36. Cronjager L, Meister D (1992) Machining of fibre and particle-reinforced aluminum. Annals CIRP 41(1):63–66
37. Coelho RT, Yamada, S, IeRoux, T, Aspinwall, DK, Wis, MLH (1993) Conventional machining of an aluminium based SIC reinforced metal matrix composite (MMC) alloy. Proceeding of the 30th MATADOR, Manchester, pp 125–133
38. Coelho RT, Aspinwall DK, Wise MLH (1994) Drilling and reaming aluminium-based Metal Matrix Composites (MMC) using PCD tooling. Trans NAMRI/SME
39. Lane C (1990) Machining characteristics of particulate-reinforced aluminum. In: Fabrication of particle reinforced metal composites. ASM, Ohio
40. Jawaid A, Barnes S, Ghadimzadeh SR (1992) Drilling of particulate aluminum silicon carbide metal matrix composites. Proceeding of the ASM Mater Workshop, Chicago, IL
41. Venkatesh VC et al. (2001) Micro-drilling of composite. In: International Conferrence on Material for Advanced Technologies. Sun Tech City, Singapore
42. Narutaki N (1996) Mach MMC's. Vdi berichte 1276:359–370
43. Songmene V, Balazinski M (1999) Machinability of graphitic metal matrix composites as a function of reinforcing particles. CIRP Annals—Manuf Tech 48(1):77
44. Songmene V, Stephenson TF, Waner AEM (1997) Machinability of Graphitic Silicon Carbide Aluminum Metal Matrix Composite GrA-NiTM. ASME Int. Mech. Eng. Congr. and Expo. Dallas, Texas
45. Tosun G, Muratoglu M (2004) The drilling of an Al/SiCp metal-matrix composites. Part I: Microstructure. Comp Sci Technol 64(2):299–308
46. Basavarajappa S et al (2007) Drilling of hybrid metal matrix composites—workpiece surface integrity. Int J Mach Tools Manuf 47(1):92
47. Lane C, Finn M (1992) Observations on using CVD diamond in milling MMCs. In: Machining and the physics of the machining process. TMS, Chicago, IL
48. Chandrasekaran H, Johansson J (1996) On the behaviour of fibre/particle reinforced aluminium alloy matrix composites in milling and grinding. VDI Berichte 1276:463–478

Chapter 5
Grinding of Metal Matrix Composites

B. Anand Ronald, L. Vijayaraghavan and R. Krishnamurthy

Abstract The metal-matrix composites are difficult-to-machine materials, since the matrix and reinforcement possess widely-different properties. The manufacturing methodology adopted has significant effect on material and product performance. This chapter presents details on the influence of reinforcement size, grinding abrasive material and grit size on the grindability of Al/SiC_p metal-matrix composites. Different performance indicators namely grinding force, temperature and acoustic emission, along with surface texture of the ground surface and chip morphology are discussed.

5.1 Introduction

Intense research in material science has been directed toward the development of new light-weight engineering materials possessing high-specific strength and stiffness at elevated temperatures combined with good creep, fatigue and wear resistance. Advanced automotive and aerospace technologies require such material specification for enhanced performance. These properties are not achievable with light-weight monolithic titanium, aluminum and magnesium alloys. Hence designers resort to composite materials such as Fiber-Reinforced Plastics (FRP),

B. A. Ronald (✉)
Department of Mechanical Engineering, SSN College of Engineering, Kalavakkam 603110, India
e-mail: anandronald@rediffmail.com

L. Vijayaraghavan · R. Krishnamurthy
Department of Mechanical Engineering, Indian Institute of Technology Madras, Chennai 600036, India
e-mail: lvijay@iitm.ac.in

R. Krishnamurthy
e-mail: ramalingamkrishnamurthy@gmail.com

J. P. Davim (ed.), *Machining of Metal Matrix Composites*,
DOI: 10.1007/978-0-85729-938-3_5,

Metal-Matrix Composites (MMC) and Ceramic-Matrix Composites (CMC). Metal-matrix composites are applicable to relatively higher load cum temperature applications owing to their strength and high stiffness-to-density ratio. The density of most MMC's is approximately one-third that of steel, resulting in high-specific strength and stiffness.

MMC's find application in different industries like aerospace (for fuselage of space shuttle orbiter and vertical tail section of advanced fighter planes), automobile engine parts (piston, cylinder liners and brake drums), sports equipments and marine parts. Especially aluminum alloys reinforced with silicon carbide are relatively new, potentially useful structural materials with high-specific strength and modulus values [1]. Polymer reinforced with SiC particle known as Epoxy—concrete has found application in sliding contact pair; however, their usage is limited to moderate working environment.

5.1.1 Difference Between Machining Conventional Materials and Composites

As in the case of polymeric composites, MMC's also pose problems during fabrication and further processing. The presence of hard and abrasive reinforcement in the metal-matrix inflicts considerable wear of tool, posing several machining problems. There are significant differences between the machining of conventional metals and alloys, and that of composite materials, because composites are normally anisotropic, inhomogeneous and reinforcement is highly abrasive in nature. Due to these material properties, machining often leads to damage being introduced in work-piece and rapid wear development in cutting tool. The high tool wear is caused by very hard and abrasive reinforcement, which comes in direct contact with the tool. Also the tool encounters alternatively matrix and reinforcement material, causing fluctuating tool load and consequent fatigue failure. This imposes special demand on the tool material specifications. The high tool wear leads to uneconomical production, or it even makes the process impossible. During machining, material separation is accomplished by brittle fracture rather than plastic deformation.

5.1.2 Need for Grinding

Although the primary processing such as in situ MMC development [2] and thixotropic processing [3] can be used to produce near-net shape components, a process like grinding is still required to produce MMC components to the desired final dimensions, with good surface finish and a damage constrained surface. Improper tooling and machining conditions not only escalate the component cost, but also cause sub-surface damage in MMC components. Such damage is detrimental to surface of critical engineering components that are designed for creep, fatigue, high-electrical frequency, etc., and is responsible for large scattering of test data.

5.1.3 Challenges in Grinding of Metal-Matrix Composites

During grinding of soft material like Aluminum, there is a tendency of chip to clog the wheel; however, the presence of reinforcement can improve the machinability in terms of both surface roughness and lower-order tendency to clog the wheel. Yet, grinding being a highly specific energy process, the increase in temperature during grinding, increases the ductility of the work-piece, thereby enhances the wheel loading. Wheel loading adds to the problem by inducing vibration; consequent surface roughening and reduced material removal rate. This calls for a case-specific wheel specification in terms of structure and bond [4]. Attempts using open structure wheel to contain the problem of wheel loading have not yielded results, owing to the selection of suitable grit. Mostly wheel specification is compromised to meet the desired performance.

5.1.4 Choice of the Grinding Wheel

Among the engineering materials, ferrous alloys/steels find maximum application in grinding. In fact grinding is an abrasive process, calls for selection of abrasive based on Mohr's scale of abrasive index. Mostly steels occupy fifth and sixth place in the scale of abrasive index, calling for materials in seventh place onwards for abrasion. Apart from that, chemical compatibility and chemical stability also play significant role in abrasive selection. Thus steels can be ground/abraded by seventh group abrasive, such as Al_2O_3, SiC and superabrasive cBN barring the hardest diamond. Unlike the case of steel, aluminum-matrix-based MMC poses challenges due to wheel loading. Also the reinforced abrasive can influence the performance of the abrasive grit by way of attrition wear, chemical interaction and abrasive dulling. Hence grinding performance can be material specific.

Among the traditional abrasives both alumina and SiC cannot be used owing to compatibility problems. Superabrasives such as diamond and cubic boron nitride could be used. Cubic boron nitride performs machining with higher order force, while diamond can lead to problem of chemistry such as material interaction and dissociation. Thus metal-matrix composites containing Aluminum matrix and reinforced SiC particles mostly facilitate mostly stochastic machining performance.

Fig. 5.1 View of the experimental set-up

5.1.5 Criteria for Grindability Evaluation

The term "*Grindability*" is defined as the ease with which a material can be ground to specification/criteria when manufacturing a finished product from raw material. Several criteria are used to evaluate grindability; the criteria may include minimum tool wear (better G-ratio), longer wheel life, lower grinding forces, temperature, vibration, better surface finish, minimum residual stress, absence of surface defects, maximum dimensional conformity, higher metal removal rate, etc. In other words grindability can be evaluated in terms of function-related integrity specification of the ground part. Unlike the case of traditional machining process, wherein machinability is related to the free machining steel (SAE 1010), no such reference material is in vogue for grindability evaluation. Hence mostly grindability can be related to user-specified criterion.

5.2 Grindability Studies on Metal Matrix Composites

Grinding trials on Al2124/SiC_p (Particle size—5 and 50 μm, with volume 30%) metal-matrix composite were carried out using silicon carbide (SiC), cBN and diamond wheels of identical grit (ASTM–80/100 mesh) and bond (resin). The influence of grit size was studied by comparing 80/100, 120/140 and 170/200 diamond wheels. Surface grinding trials have been carried out on a tool and cutter grinding machine.

Figure 5.1 shows the view of the experimental set-up and the experimental details are presented in Table 5.1. The diameter of the wheel was 150 mm. The performance of the grinding wheel was assessed in terms of grinding force (using dynamometer), temperature (using Infra Red Pyrometer) and acoustic emission (AE) signal characteristics from the work-piece.

Table 5.1 Grinding conditions

Parameter	Range
Wheel speed (V)	1,000–2,800 m/min
Feed (s)	0.2–0.9 m/min
Depth of cut (a)	10–30 μm

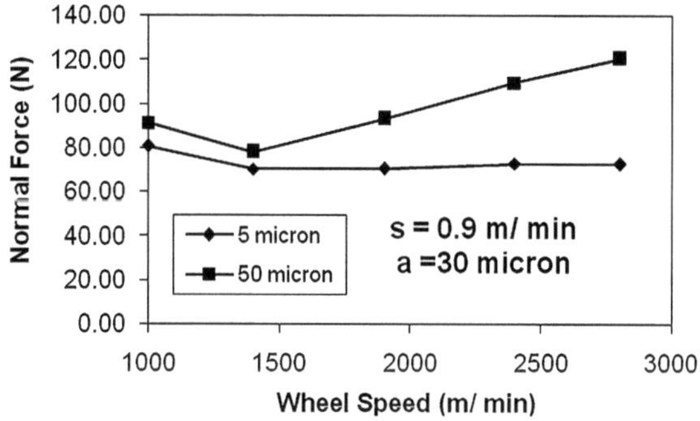

Fig. 5.2 Typical parametric influence on normal force with grinding condition for 5 and 50 μm SiC_p reinforced MMC

5.2.1 Grinding Force: Parametric Influence

The Al/SiC_p composite specimen was mounted on a piezoelectric grinding dynamometer for monitoring the force components. It is to be noted that monitoring the force components in a dynamic process such as grinding calls for piezoelectric transducers for reliable monitoring.

5.2.1.1 Influence of Size of the Reinforcement

The structure, property and functional relationship of metal-matrix composite are influenced by the matrix material and the reinforcement in terms of its chemistry, size and volume fraction. Grindability of MMC's with reinforcement of two different sizes is evaluated by monitoring the force component. Typical monitored variation of normal force component with grinding condition is shown in Fig. 5.2.

It is seen that MMC with coarser reinforcement exhibits relatively higher order force component. The composite exhibits higher order force especially after wheel speed of 1,400 m/min, while MMC with finer reinforcement is relatively insensitive to wheel speed as for as normal force is concerned. The observed increase in normal force component with higher grinding speed can be attributed to possible interaction between diamond and SiC, facilitating graphitization of diamond and consequent rise in force component. The parametric influence on tangential force component also indicates a similar trend as illustrated in Fig. 5.3.

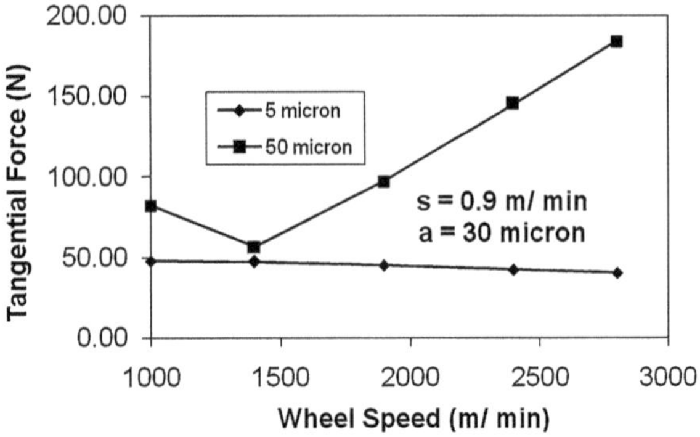

Fig. 5.3 Typical parametric influence of tangential force with grinding condition for 5 and 50 μm SiC_p reinforced MMC

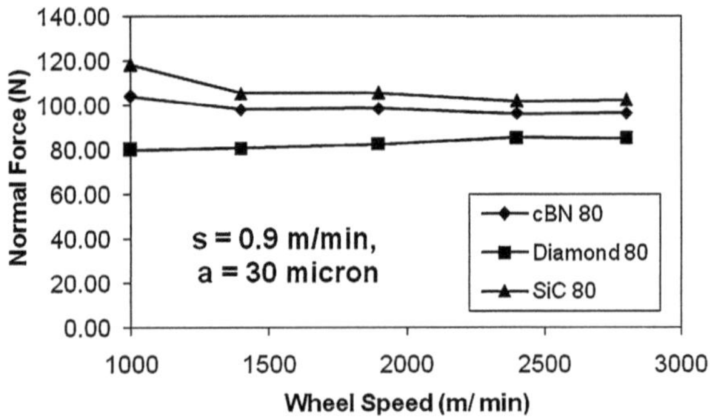

Fig. 5.4 Typical parametric influence on normal force for different grit material

5.2.1.2 Influence of Type of Grit

Typical parametric influence on normal force during grinding with different grit material is shown in Fig. 5.4. Observation on normal force component indicates that, SiC wheel experiences higher order force compared to cBN and diamond wheel. Diamond being a harder abrasive with better cleavage exhibits better cutting ability than cBN and SiC, hence the grinding force is relatively less.

The SiC wheel exhibits higher order grinding force, owing to chemical compatibility with the reinforcement (SiC_p), leading to abrasive wear and sliding dominant grinding. Also based on a study on grindability of cast Aluminum composites [5], it is reported that SiC wheels have relatively less G-ratio, compared to diamond wheels. It is seen that with lower speeds, both SiC and cBN exhibit higher rate normal force component attributable to plowing. With increased speed, a reduction in normal force can be seen. However, with higher grinding speeds, the abrasive grinds with a marginal variation in normal force. The observed speed with insensitive nature of normal force could be attributed to possible loading of the wheel, induced by grinding temperature associated with higher wheel speeds. Among the abrasive grit, diamond outperforms others by way of lower order normal force component.

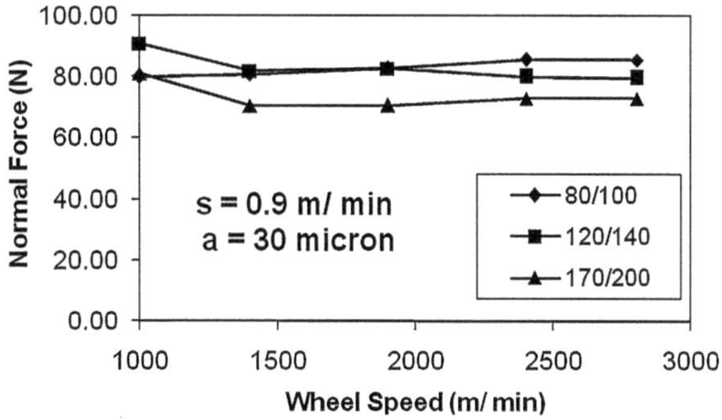

Fig. 5.5 Typical parametric influence on normal force component for diamond wheels of varying grit size

5.2.1.3 Influence of Abrasive Grit Size

Apart from the type of grit, grit size also influences grindability of MMCs. Typically observed influence of grit size (diamond) on grinding force component is shown in Fig. 5.5. It is seen that coarser grit grinds Al/SiC_p composite with relatively higher order force component. Increased tendency to loading of wheel with finer grit accounts for the reduction in force component.

5.2.2 Observation on Grinding Temperature

Grinding is usually a high-speed process, associated with higher order temperature. Grinding temperature comprises of cutting temperature (θ_c) and sliding temperature (θ_s); i.e., $\theta_{grinding} = \theta_c + \theta_s$. The temperature in cutting θ_c can be expressed as $\theta_c \sim V^b$,

where
V Cutting speed (in m/min)
b Material-dependent exponent

Sliding temperature θ_s is due to sliding of the abrasive on the groove made by the leading abrasive. This can be minimized by proper selection of the grinding fluid. The grinding fluid selected has to carry/dissipate the cutting temperature so as to adequately minimize the sliding component. This requires a balance/compromise in its selection.

In any abrasive process mechanical, thermal and even chemical effects are usually superimposed in the contact zone. Every material removal process, especially abrasive process, generates significant amount of heat, which can lead to deviation in the dimensional accuracy of the work-piece, an undesirable change of the surface integrity state due to relatively hostile environment. Grinding processes transform kinetic energy mainly into thermal energy. A substantial part of the thermal energy is dissipated (partition ratio) into the work-piece and heats up the work-piece surface and sub-surface layer. This may lead to sub-surface damages

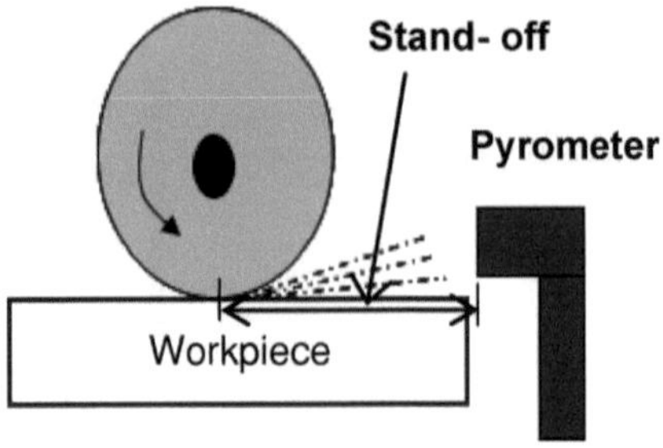

Fig. 5.6 Schematic diagram of the temperature measurement

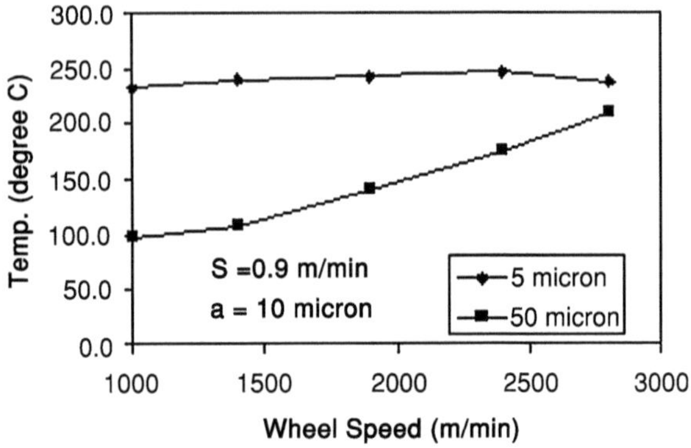

Fig. 5.7 Typical parametric influence on grinding temperature for 5 and 50 μm SiC_p reinforced MMC

like phase transformation, hardness alterations, surfacial residual stress and in extreme cases surface and sub-surface cracking. One method to avoid detrimental influences is to measure the surface temperature of the work-piece and carry out the required corrective action to maintain the process status.

Although it may not be easy to assess the exact temperature in the grinding zone; the trend in variation of grinding temperature can be evaluated by focusing an infrared radiation type pyrometer on the chip trace (Fig. 5.6).

5.2.2.1 Influence of Size of the Reinforcement

Typical monitored parametric influence on temperature measured on spark stream (referred to as grinding temperature) is shown in Fig. 5.7. It is seen that grinding of MMC with finer reinforcement particles is associated with higher order temperature. Also it is mostly insensitive to wheel speed. Coarser reinforcement in MMC has resulted in reduced order of temperature; also with wheel speed, it tends to rise.

5.2.2.2 Influence of Type of Grit

Typical illustration on variation of grinding temperature with grinding conditions for the different wheels is shown in Fig. 5.8. There is rise in grinding temperature with grinding conditions. The diamond wheel experiences relatively less

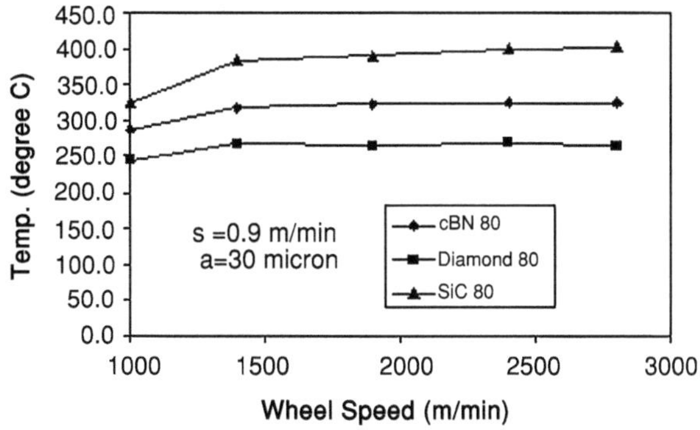

Fig. 5.8 Typical parametric influence on grinding temperature for different grit material

temperature compared to the cBN and SiC wheel. Referring to the illustration on grinding temperature, it can be seen that while there is a progressive rise in grinding temperature with wheel speed, the trend of variation is seen to change above 1,400 m/min grinding speed. This indicates that irrespective of the wheel, a critical speed of 1,400 m/min, can be seen for grinding of Al/SiC$_p$ composite.

It can be seen that with higher grinding condition, the temperature is speed insensitive, indicating possible degradation in the performance of the grinding wheel and also thermal softening of the matrix material. Also from the previous literature [6], we understand that there is a possibility of degradation of diamond, especially at higher temperatures and also due to interaction with the SiC$_p$ reinforcement. Among the three wheels, SiC wheel exhibits a higher order temperature indicating relatively poor performance.

5.2.2.3 Influence of Abrasive Grit Size

During grinding with finer grit diamond wheel, there is a greater tendency of wheel loading and hence rubbing, which was reflected in the temperature during grinding. With loading, the sliding component of grinding temperature increases, accounting for the higher order temperature [7].

5.2.3 Material Response: Chip Morphology

Grinding is largely carried out on hard, high-strength ferrous materials; relatively softer non-ferrous material is usually not an ideal material for abrasion. Hence composite materials of a non-ferrous material matrix (like aluminum) can pose problems in processing due to the presence of harder reinforcement particles in relatively softer metal-matrix [8]. During grinding of Al/SiC$_p$ composites, the relatively softer/ductile aluminum encountered by the abrasive of higher negative rake (negative cutting wedge) angle, will experience upsetting ahead of the

Fig. 5.9 SEM macrograph of chips for $V = 1{,}000$ m/min, $s = 0.2$ m/min and $a = 10$ μm [Diamond Wheel]

Fig. 5.10 SEM macrograph of chips for $V = 1{,}000$ m/min, $s = 0.9$ m/min and $a = 30$ μm [Diamond Wheel]

cutting wedge and due to continuous encounter, the upset lump would be displaced as a lumped/upset mass in the chip (by the interfacial adherence of successive lumps) [9].

The interaction between a negative-wedged tool (abrasive) and work material, results in creation of stagnation zone and subsequent upsetting of work material. The flow of material around the negative wedge would be partly due to rubbing and sliding under the tool, followed by flow of upset material. This results in a highly-strained material, moving as a lumped mass of chip [10]. With finer depth of grinding and feed rate the ductile-aluminum matrix will experience brittle mode of machining, leading to production of short segmental/powdery chip (Fig. 5.9).

With higher order depth of grinding and feed rate, the material experiences less upsetting resulting in elongated streaks of chip (Fig. 5.10).

With increased speed and moderate depth of grinding, relatively longer streaks of chip occur (Fig. 5.11).

Unlike the case of diamond wheel, grinding with SiC wheel has resulted in higher order temperature (Fig. 5.8) and consequent ductile mode of chip production. Relatively larger ductile chip can be seen in Fig. 5.12.

Fig. 5.11 SEM macrograph of chips for $V = 1{,}400$ m/min, $s = 0.9$ m/min and $a = 20$ μm [Diamond Wheel]

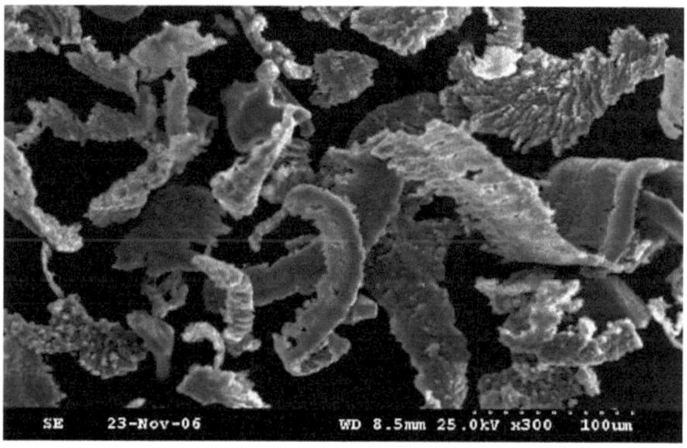

Fig. 5.12 SEM macrograph of chips for $V = 2{,}800$ m/min, $s = 0.2$ m/min, $a = 30$ μm [SiC Wheel]

5.2.4 Acoustic Emission Monitoring

Acoustic emission is defined as the transient elastic stress wave due to the rapid release of energy from a material, when subjected to a state of stress [11]. This energy release is associated with the abrupt redistribution of internal stresses and as a result of this, a stress wave (locked in elastic stress) is propagated through the material. The definition of AE given above indicates that, processes that are capable of changing the internal structure of a material such as generation of dislocation, dislocation motion, directional diffusion, creep, grain boundary sliding and twinning, associated with plastic deformation, phase transformations, vacancy coalescence and decohesion of inclusions, usually associated with temperature and fracture are sources of acoustic emission. Of the processes said above, only plastic deformation and fracture are of significance in metal cutting, while temperature-associated events can also occur in grinding. Since AE is emitted by a material under a state of stress, it is established as an effective tool for monitoring of machining process/status of material [12].

Also grinding is relatively a high cutting speed process, associated with higher order strain rate and temperature. i.e., the work material undergoes localized

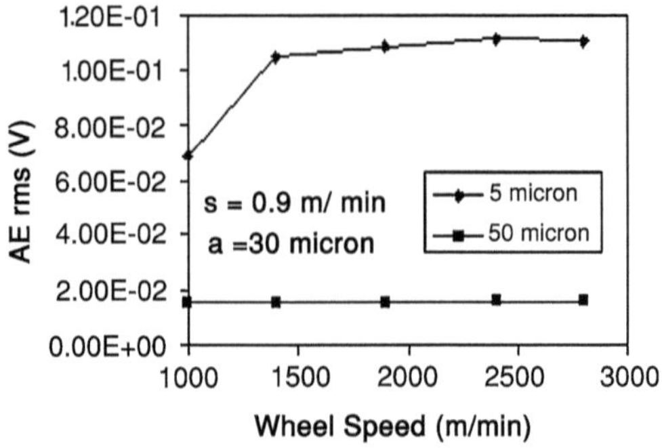

Fig. 5.13 Typical parametric influence on AE_{rms} for 5 and 50 μm SiC_p reinforced MMC

deformation and concurrent heating. This will influence the elastic stress energy release. In addition, during grinding of Al/SiC_p composites one can anticipate plowing, sliding and smearing of softer aluminum matrix, and pull-out/dislodgement of reinforcement (SiC_p). This will result in continuous changes in the wheel–work interface condition and consequent material response in terms of acoustic emission.

The response of the work material to grinding environment has been assessed by monitoring the acoustic emission from the work-piece during grinding. The piezoelectric AE sensor was mounted in such a way, to ensure minimum signal attenuation. The bottom portion of the work-piece was made to be in acoustic contact with AE sensor, by means of a couplant. Acoustic emission is normally a low amplitude, high-frequency elastic stress (energy) wave released from a material under stress. Hence monitoring of AE signal from the work-piece can be a realistic way of assessing the material response during grinding.

5.2.4.1 Influence of Size of SiC Reinforcement on AE_{rms}

Typical monitored influence of size of the reinforced particle on 'rms' of the acoustic emission from the work-piece during grinding is illustrated in Fig. 5.13.

It is seen that with finer reinforcement, more acoustic activity occurs. Also with higher speeds, especially with finer reinforcement, more or less steady acoustic activity occurs. Such occurrences are attributed to wheel loading, associated with higher grinding temperature.

5.2.4.2 Influence of Type of Grit

Typical assessed power spectra of the AE signal monitored during grinding with diamond, cBN and SiC wheel are illustrated in Fig. 5.14. The power spectrum contains dominant peak mostly around 50 kHz and 150–200 kHz frequency range. The signal is mostly of mixed mode, while the low-frequency peaks indicate continuous/deformation mode of emission, the dominant high-frequency peaks are attributed to burst mode of emission, i.e., during grinding of Al/SiC_p metal matrix

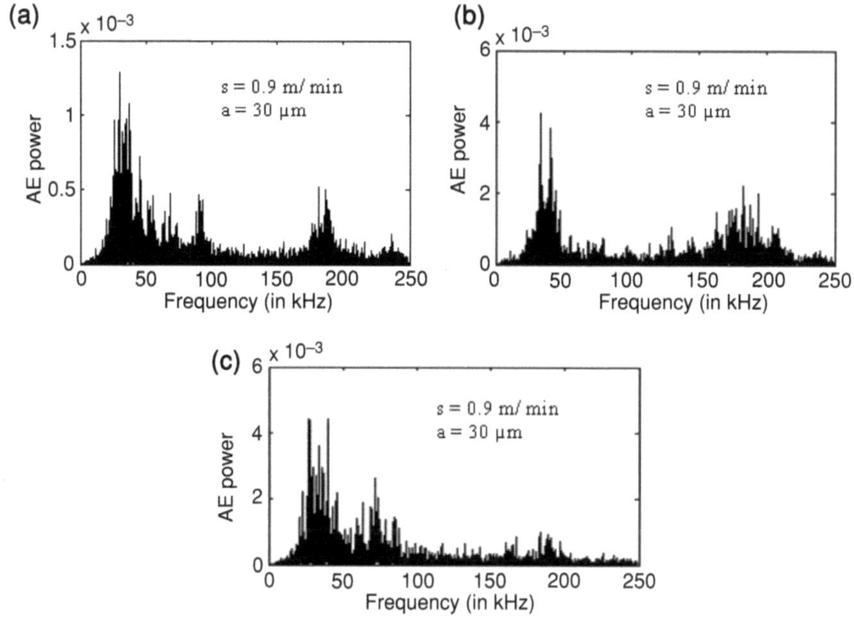

Fig. 5.14 Typical AE power spectrum plots for the three grinding wheels. **a** Diamond wheel. **b** cBN wheel. **c** SiC wheel

composites, the grinding will be associated with plowing (softer matrix), upsetting of the matrix, dislodgement of the upset material as segmental chip. The production of continuous-lumped chips results in continuous mode of emission, while spalling of the loaded material, dislodging of reinforced SiC_p and even chipping of diamond grit contribute to the burst mode.

From the illustration for diamond wheel (Fig. 5.14a), the power spectra shows relatively lower peak power, when compared with AE response between cBN and SiC wheel for similar grinding conditions. Unlike the case of diamond wheel, the power spectrum for cBN wheel (Fig. 5.14b) contains mostly dominant low-frequency peaks and higher power. This is indicative of continuous mode of emission and can be attributed to sliding dominant grinding in the case of cBN wheels. Unlike diamond, which retains its sharpness due to inherent cleavage activity, cBN normally deforms [13] when exposed to temperature and pressure causing a rise in force and tendency to slide. This is reflected in the observed higher order normal force component (Fig. 5.4) and grinding temperature (Fig. 5.8) compared to diamond grit. The AE power spectrum for the SiC wheel (Fig. 5.14c) is mostly of relatively higher power, compared to the other two wheels, for similar grinding conditions. Also more number of low-frequency peaks can be observed, indicative of the sliding dominant grinding, possibly due to attrition wear of SiC abrasive.

Typical AE_{rms} values are shown in Fig. 5.15. The AE_{rms} value tends to rise with wheel speed for all three wheels. It is to be noted that despite the reduction

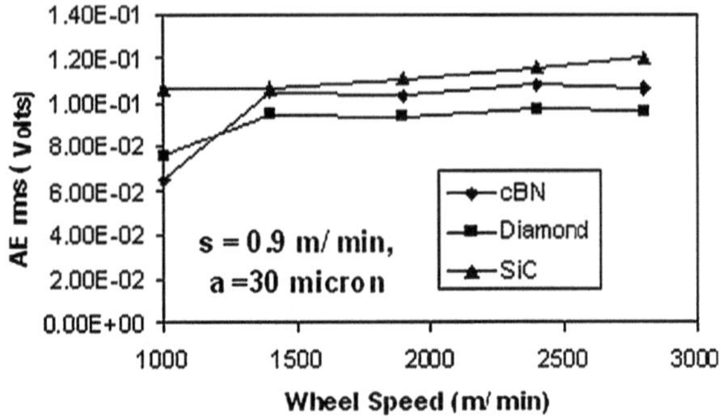

Fig. 5.15 Typical AE_{rms} values for the three wheels

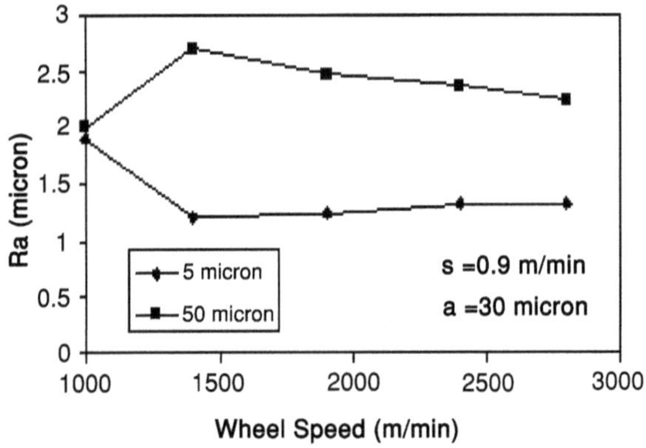

Fig. 5.16 Typical parametric influence on surface finish for 5 and 50 μm SiC_p reinforced MMC

observed in normal force components, AE_{rms} increases with wheel speed. The thermal softening of the work material could have resulted in higher energy release rate, which is supplemented by the observed rise in grinding temperature (Fig. 5.8).

5.2.5 Observation on Surface Finish

Advanced machining processes are becoming increasingly productive, leading to an increasing concentration of energy in the material removal area. As a consequence, the generated surfaces are significantly influenced by the thermo-mechanical loadings encountered during the cutting or abrasive processing. The quality of the surface influences the component performance by way of tribological response. An important measure of grindability is surface finish. The surface finish is influenced by grinding condition, the nature of composite and the wheel specification.

Typical monitored parametric influence on surface roughness (Ra) is shown in Fig. 5.16. It is seen that with finer reinforcement, it is possible to attain good surface finish; also with grinding wheel speed, an improved texture is attainable.

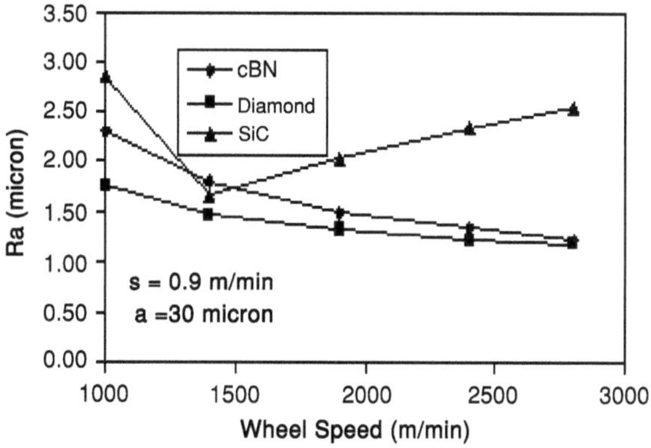

Fig. 5.17 Typical parametric influence on surface finish for different grit material

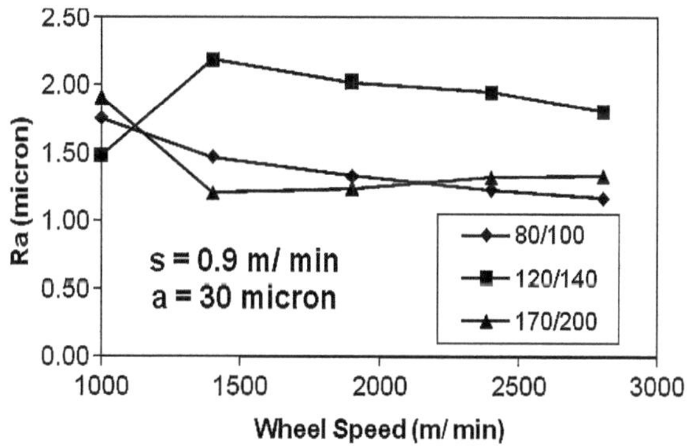

Fig. 5.18 Typical parametric influence on surface finish for diamond wheels of varying grit size

The significance of type of abrasive/grit used for grinding, on surface roughness is illustrated in Fig. 5.17. It is seen that with SiC, a roughened texture results, while with diamond grit relatively finer/smooth texture is attainable. It is seen that above a certain speed, the surface texture tends to be rougher, possibly accounting for the occurrence of wheel loading and even possible degradation of the abrasive.

From the illustration on surface finish in Fig. 5.17, it can be observed that diamond wheel gives better surface finish compared to cBN and SiC wheels. This is due to the better grindability of Al/SiC$_p$ metal-matrix composites with diamond wheel, in terms of lower order grinding force, temperature and acoustic emission.

With higher speed, only a marginal change in surface finish can be seen. This is supplemented by the observed rise in grinding temperature beyond 1,400 m/min of speed, inducing thermal softening of the work material and loading associated grinding. At higher wheel speeds, due to higher grinding temperatures there is a possibility of graphitization of diamond (partial grinding) and thereby leading to deterioration in the grinding performance. For the SiC wheel the surface finish also deteriorates appreciably at higher grinding condition, probably due to chemical compatibility with SiC$_p$ in the composite. The cBN wheel also gives rougher surface texture, compared to diamond wheel. This improvement in surface finish

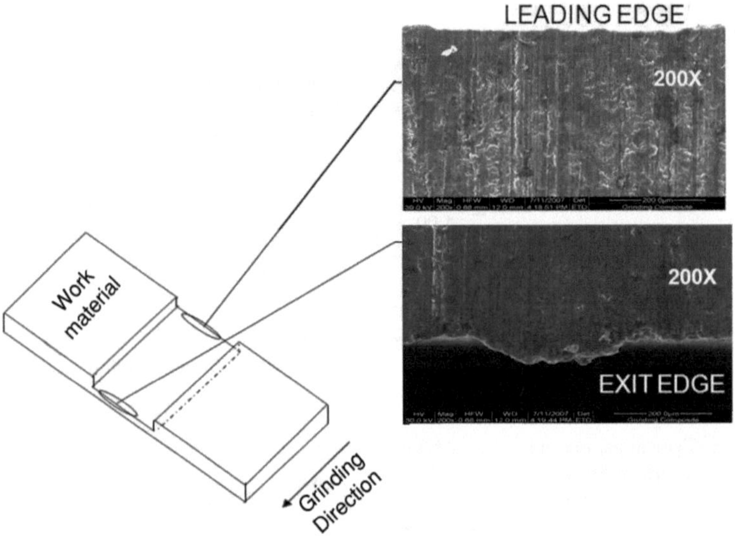

Fig. 5.19 Macrograph of a zone over the leading and exit edges of the work material

with wheel speeds is mainly due to abusive grinding (rubbing), rather than abrasive grinding.

From the observation on surface finish of ground MMC, it is seen that SiC grit cannot be chosen from the point of view of surface texture/finish. Also among cBN and diamond grits, while cBN tends to maintain consistent trend of variation in surface finish with grinding condition, diamond grit exhibits a tendency to produce rougher texture with higher wheel speeds. However, comparatively, diamond grit maintains better surface texture.

Significance of grit size (abrasive) on surface texture is presented herein. Typical monitored variation of surface texture (Ra) with grinding condition, as influenced by the grit size is illustrated in Fig. 5.18.

It is seen that with higher speed, only marginal variation in surface roughness occurs. This is mostly attributable to the loading of the grinding wheel with higher speed/temperature. Also grinding with finer grit facilitates good surface texture.

5.2.6 Surface Texture: Macrograph

During grinding of Al/SiC_p metal-matrix composites, one can anticipate smearing of softer aluminum matrix and pull-out/dislodgement of reinforcement (SiC_p) from the work material. Simultaneously loading of the wheel and spalling of loaded material can take place in the wheel. Typical macrograph of the leading and exit edge of the work material after grinding is shown in Fig. 5.19. In the leading edge

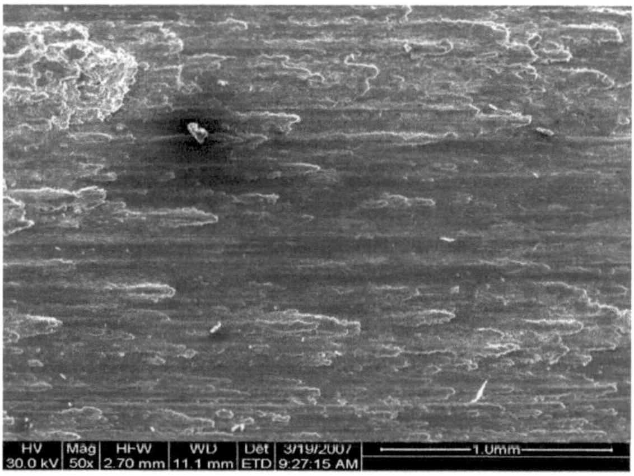

Fig. 5.20 Ground surface texture for grinding condition $V = 2{,}800$ m/min, $s = 0.9$ m/min, $a = 30$ μm [Diamond 80/100 Wheel]

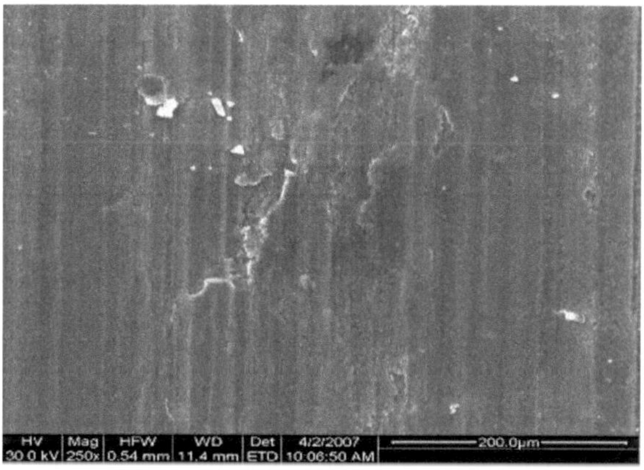

Fig. 5.21 Ground surface texture for grinding condition $V = 1{,}000$ m/min, $s = 0.2$ m/min, $a = 10$ μm [cBN Wheel]

there is no smearing, whereas smearing of the work material is present on the exit edge i.e., the matrix material is deformed and pushed along the grinding direction forming a ledge over the edge; also the surface texture in the exit region is partially glazed (absence of grinding lay).

Typical surface texture of Al/SiC$_p$ metal-matrix composites ground with diamond wheel is presented in Fig. 5.20. With higher grinding conditions the texture clearly shows intense flow of material, with whitening of the edges of flow pattern indicating possible oxidation of the hot material.

Due to intense adhesion of the work material on to the wheel surface, and consequent squeezing, there is possibility of pull-out of surface material by mutual transfer and also smearing of the softer matrix on the work material.

The surface ground with cBN wheel shows pull-out of reinforcement, at lower (Fig. 5.21) and higher (Fig. 5.22) grinding conditions. From the literature, it can

Fig. 5.22 Ground surface texture for grinding condition $V = 2{,}800$ m/min, $s = 0.9$ m/min, $a = 30$ μm [cBN Wheel]

Fig. 5.23 Ground surface texture for grinding condition $V = 1{,}400$ m/min, $s = 0.6$ m/min, $a = 20$ μm [SiC wheel]

be inferred that fracture, pull-out of the reinforcement particle is practically unavoidable in metal-matrix composites [8].

At higher wheel speeds (Fig. 5.22), mostly glazed texture can be seen, along with particle pull-out. This is indicated by the matte texture seen in the figure.

From the observation (Fig. 5.23) of surface texture pertaining to grinding with SiC wheel at relatively lower grinding conditions (speed, feed, and grinding depth) the ground surface presents a matte texture possibly due to surface melting/cooling. The presence of molten-solidified globules on the surface is indicative of the higher order grinding temperature.

With finer grit diamond wheel, at higher grinding conditions (Fig. 5.24), the ground surface shows relatively rougher texture, along with molten-spherical globules. This indicates that the temperature during grinding is sufficiently high to melt the matrix material and supplements the observation, as to why the finer grit wheel could not yield better surface finish. Whereas such molten globule

Fig. 5.24 Ground surface texture for grinding condition $V = 2{,}800$ m/min, $s = 0.9$ m/min, $a = 30$ μm, diamond 170/200 wheel

formation was not found on the ground surface while grinding with coarser grit diamond wheel (Fig. 5.20) for similar grinding conditions.

5.3 Conclusion

Better grindability can be achieved for MMC reinforced with finer SiC particle, compared to coarser SiC_p reinforced MMC. From the above discussion on the influence of wheel material on grinding performance, diamond wheel out-performs other wheels, by way of lower order grinding force, temperature and AE energy release rate, i.e., diamond can grind MMC's with relatively lower temperature compared to other wheels thereby giving better structural integrity. Also irrespective of the wheel, a critical speed of 1,400 m/min has been observed, possibly due to temperature dependence of the matrix material. The presence of SiC reinforcement makes SiC wheel unsuitable for grinding, owing to possible attrition wear of abrasive grit. With higher order grinding condition, the interaction between SiC and diamond grit results in the instability of diamond. Despite the relatively higher order force components, the performance of coarser grit diamond wheel can be reckoned with, owing to minimal loading. Usually the finer grit wheel gives better finish, but in the case of MMC, due to proneness to loading (softer matrix material) gives poor performance.

References

1. Zhong Z, Hung NP (2002) Grinding of alumina/aluminium composites. J Mater Process Technol 123(1):13–17

2. Hu C, Barnard L, Mridha S, Baker TN (1996) The role of SiC particulate and Al_2O_3 (saffil), fibers in several alloys during the formation of in situ MMCs developed by laser processing. J Mater Process Technol 58:87–95
3. L Shy-Wen, Chung DDL (1996) Shaping metal-matrix composites by thixotropic machining. J Mater Process Technol 58:67–69
4. Anand Ronald B, Vijayaraghavan L, Krishnamurthy R (2009) Studies on the influence of grinding wheel bond material on the grindability of metal matrix composites. Mater Des 30(3):679–686
5. Vaccari JA, Lane CT (1993) Machining a new breed of aluminium. Am Mach Nov 37(11):56–60
6. Beffort O, Vaucher S, Khalid FA (2004) On the thermal and chemical stability of diamond during processing of Al/diamond composites by liquid metal infiltration (squeeze casting). Diam Relat Mater 13:1834–1843
7. Anand Ronald B, Vijayaraghavan L, Krishnamurthy R (2009) Grinding of Al/SiC_p metal matrix composites, with diamond wheels of varying grit. J Mach Form Technol 1(3/4): 221–236
8. Teti R (2002) Machining of composite materials. Ann CIRP 51(2):611–634
9. Komanduri R (1971) Some aspects of machining with negative rake tools simulating grinding. Int J Mach Tool Des Res. 11:223–233
10. Anand Ronald B, Vijayaraghavan L, Krishnamurthy R (2010) Chip morphology—an indicator of response of composite material to grinding. Int J Mater Prod Technol 37(1/2): 71–82 (special issue on “Experimental and Numerical Studies on Machinability of Materials”)
11. ASTM (1982) E 610-Standard definition of terms relating to acoustic emission. Am Soc Test Mater, Philadelphia, USA
12. Inasaki I (1998) Application of acoustic emission sensor for monitoring machining processes. Ultrasonics 36:273–281
13. Sreejith PS, Krishnamurthy R, Malhotra SK (2007) Effect of specific cutting pressure and temperature during machining of carbon/phenolic ablative composite using PCBN tools. J Mater Process Technol 183:88–95

Chapter 6
Dry Cutting of SiC Particulates Reinforced Metal Matrix Composite

Arnaud Kremer and Mohamed El Mansori

This chapter addresses the cutting of Metal Matrix Composite with SiC particulate reinforcement (MMCp) to highlight the physical mechanisms that govern the material removal in dry mode. The performance of a variety of polycrystalline diamond (PCD) tools and nanostructured diamond coatings are studied when cutting MMCp with different densities of SiC particulate reinforcement. The simultaneous occurrence of adhesive wear mode and interface consumption by abrasion, increases the role of the cutting tool structure (homogeneity, multiple interfaces, etc.). The use of an environmental criterion to rate the MMCp machinability shows that the process of dust emission is strongly related to tool behavior and the predominance of friction phenomena that arise at the tool/chip interface.

6.1 Introduction

Metal Matrix Composite (MMC) investigated in this chapter was obtained by powder metallurgy process followed by hot extrusion and a T4 heat treatment (solution heat treatment, water quenching and naturally aging). This treatment is described in Fig. 6.1.

Various matrices and levels of reinforcement were used. As shown in the Fig. 6.2, the SiC (silicon carbide) polyhedral particles whose largest dimension

A. Kremer (✉)
LMPF, EA 4106, Arts et Metiers ParisTech, rue Saint Dominique,
51000 Chalons-en-Champagne, France
e-mail: arnaud.kremer@ensam.eu

M. El Mansori
Mecasurf, EA 4496, Arts et Metiers ParisTech, cours des Arts et Metiers,
13628 Aix-en-Provence, France
e-mail: mohamed.elmansori@ensam.eu

J. P. Davim (ed.), *Machining of Metal Matrix Composites*,
DOI: 10.1007/978-0-85729-938-3_6,

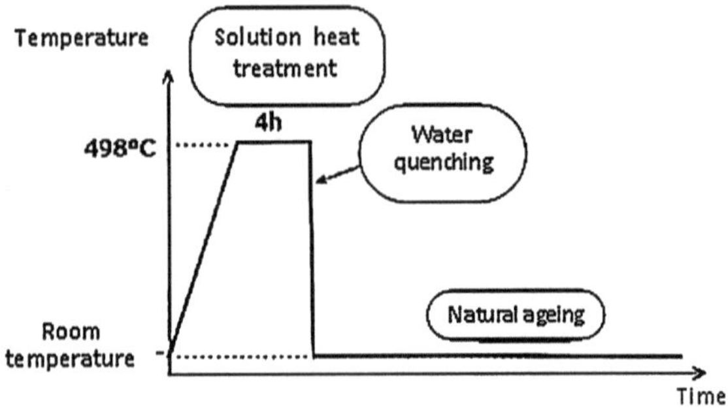

Fig. 6.1 Microstructure of the Al/15% SiC MMC

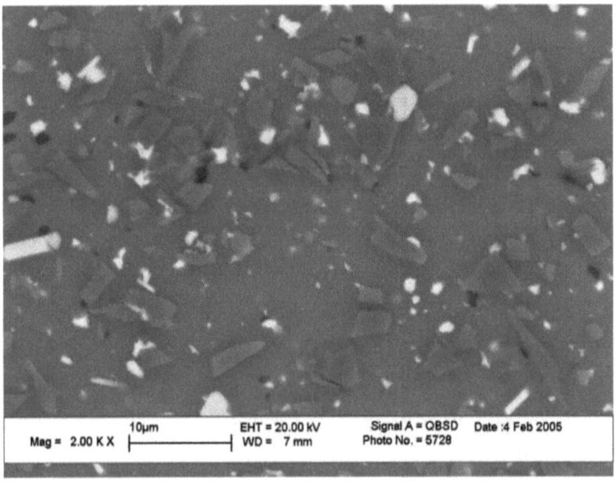

Fig. 6.2 Microstructure of the Al/15% SiC MMC

Table 6.1 Matrices composition

Aluminum alloy 2009 (matrix of CMM_0, CMM_{15}, CMM_{25})						
Cu	Mg	O	Zn	Si	Fe	Al
3.2–4.4	1–1.6	0.6	0.1	0.25	0.2	Reminder

Aluminum alloy 2124 (matrix of CMM_{35})								
Cu	Mg	Mn	Fe	Zn	Si	Ti	Other	Al
3.8–4.9	1.2–1.8	0.3–0.9	0.3	0.25	0.2	0.15	0.15	Reminder

ranges from 4 to 12 μm with a homogeneous repartition inside matrix. Four levels of reinforcement were used with the same granulometry, level rates were about 0, 15, 25, and 35%. All tested materials were non-ferrous metals where aluminum (Al) was the predominant metal and copper was the main alloying element. The chemical compositions of the two different matrices are given in Table 6.1.

For the rest of the chapter, MMCs used are referenced as CMM 0 (Al 2009—0% SiC), CMM 15 (Al 2009—15% SiC), CMM 25 (Al 2009—25% SiC), CMM 35 (Al 2124—35% SiC).

6.2 Selection of Tool Material

Machining Al/SiC composites is one of the major problems which resist its widespread engineering applications. Previous studies have shown that tool wear is excessive and the surface finish is poor when carbide tools are used. The SiC particles are much harder than the WC tool material and that leads to a high wear rate by abrasion. The cutting edge is rapidly worn, and that results to poor surface finishing. In addition, due to friction, high temperature and pressure, the Al/SiC composite work-material adheres to the cutting edge to form a built-up-edge (BUE) which also has a negative effect on surface finish [1].

Diamond tools seem to be the best tool to machine Al/SiC MMCs with acceptable tool life [2–6]. Diamond is harder than SiC and does not have chemical tendency to react with work material. For cutting tools, diamond is mainly used in two forms: either brazed PCD or chemical vapor deposition (CVD). PCD tools consist of a thin layer of fine diamond particles sintered together and brazed onto a cemented carbide substrate. CVD diamond is a more recent superhard tool material; it consists of pure diamond coating over a carbide substrate. Compared to PCD, CVD diamond is harder, exhibits a lower-friction coefficient, higher-abrasion resistance, higher-thermal conductivity and better chemical and thermal stability. Other drawbacks associated to PCD tools are the presence of cobalt binder, which limits the cutting speed and its high cost [7–9]. The high thermal conductivity of CVD diamond tools allows heat dissipation and leads to a more uniform and a reduced level temperature distribution avoiding tool failure and limiting adhesive wear. However, the main problem of CVD tools is the ability to adhere on tool tip and to resist to coating peeling [10]. In order to compare diamond tools behavior, all kind of diamond tools were used.

6.2.1 CVD Tools

The CVD cutting tools were cemented carbide tools in the form of triangular inserts coated with diamond. Four different diamond coatings obtained by CVD were tested and denoted C1, C2, C3 and C1810. These tools present different nanostructured diamond coatings. The morphology of the films was characterized by scanning electron microscope, SEM images of diamond films are shown in Fig. 6.3. In order to have a better description of the coatings, white light interferometer (WLI) was used to obtain rake face profile. Coatings C1, C2 and C3 were specially developed by Balzers Luxembourg with 6 μm thickness. Coating C1 is a monolayer coating; sharp diamond crystals can be observed in the micrograph, leading to a rough surface coating.

Diamond film C2 is also a monolayer coating, it presents a "cauliflower-like" typical morphology, leading to a low-surface roughness (smooth surface coating). Coating C3 is a combination of the two previous diamond films (smooth layer over

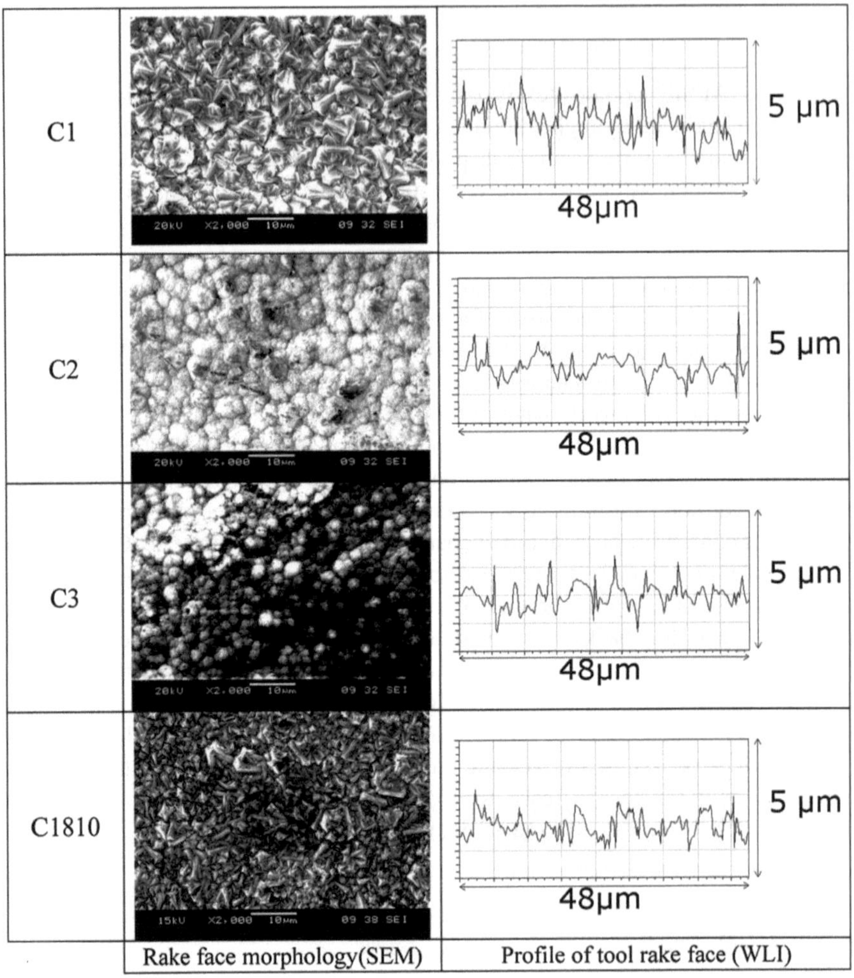

Fig. 6.3 Morphology and profile of CVD tools

a rough layer). All Balzer's coatings have the same geometry: relief angle of 7°, rake angle of 0° and a tool nose radius of 0.4 mm.

C1810 coating is a commercial grade from Sandvik Coromant. The nanostructure looks like C1 structure with smallest diamonds. Concerning the geometry, the only difference is a rake angle of 20°.

6.2.2 PCD Tools

PCD tools were provided by Asahi Diamond Industrial Europe in two grades (named PCD A and PCD B). For the first PCD grade (referenced as PCD A),

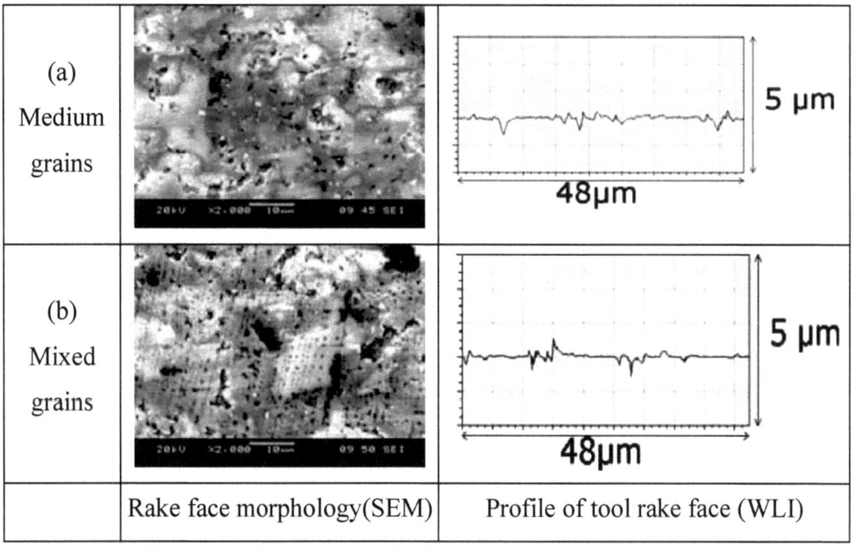

Fig. 6.4 PCD tools: PCD A (**a**), PCD B (**b**)

diamond grains of the insert have medium size, whereas diamond grains of PCB B are mixed grains (Fig. 6.4). Each PCD grade have the same geometry as CVD tools (triangular tip with clearance angle of 7° and tool nose radius of 0.4 mm) and was tested with two different rake angles (0 and 10°, referenced as PCD A0 or PCD A10, etc.).

6.3 Experimental Procedure

The wide selection of diamond tools and MMCs permit to make several experiments. Various series of tests were conducted and results were analyzed and aggregated in several subsections as:

- Characterization of machinability of MMC;
- Tool behavior and performance;
- Environmental impact of machining.

In the machining of difficult-to-machine materials, the consumption of cooling lubricant during the machining operations remains very important. The associated costs of coolant acquisition, use, its disposal and washing the machined components are significant. To reduce the manufacturing costs and to make the processes environmentally safe, the goal is to move toward dry cutting by eliminating or minimizing cutting fluids. The low-friction coefficient associated to excellent thermal properties and high-wear resistance of diamond tools makes them suitable candidates for dry machining applications [8, 9, 11–13]. Moreover during machining of MMC, insufficient lubrication could have negative effects by generating abrasive mud [14] located at the cutting edge (the mud is a melting of chips, lubricant and extracted particles). Moreover, Hung et al. [14] shown if lubricant quantity is sufficient to flush away the chip from cutting area, it does not improve tool life and surface roughness. Sutherland et al. [15] have analyzed the impact of lubrication during machining MMC on dust emission. It was demonstrated that wet turning produces much more airborne particulate matter as compared to dry turning, mainly by vaporization/condensation and atomization mechanisms. All tests were carried out under dry cutting conditions.

6.4 Characterization of Machinability

Bar turning tests were performed on a CNC RAMO RTN30 lathe using only CVD tools C1, C2 and C3. During these tests MMCs machined were CMM 5, CMM 15 and CMM 25. The aim of this section was to define the optimal cutting speed. The depth of cut was 1 mm, with cutting speeds in the range of 400–1,000 m min^{-1} and two feed rates 0.1 and 0.3 mm rev^{-1} were chosen. The cutting forces were measured using a Kistler three components piezoelectric dynamometer 9,265 B and the cutting inserts were observed after the first stage of tool wear (short duration test) using scanning electron microscopy (SEM) and optical microscopy.

6.4.1 Cutting Forces

During bar turning tests, cutting and feed forces were measured. The examination of cutting forces, Figs. 6.5 and 6.6, shows that the mean level of the component F_c appears to be the same for the three MMCs (CMM 5, 15 and 25%) and whichever coated tool used, about 110 N for the feed rate 0.1 mm rev^{-1} and 250 N for the feed rate 0.3 mm rev^{-1}. For the CMM 5 and 25, the cutting force component F_c decreases slightly with cutting speed, and for the MMC 15 it presents a minimal value for $V_c = 700$ m min^{-1}.

6.4.2 Tool Wear Mode

Figure 6.7 presents a typical SEM images of the diamond-coated tools after short-duration tests in dry cutting conditions. The transfer of material from the workpiece to the tool surfaces is clearly evident in this figure. Different zones of transferred material are identified: the first one corresponds to an accumulation of workpiece material close to the cutting edge forming a BUE. The second one corresponds to a multilayer deposition of workpiece material on the rake face and

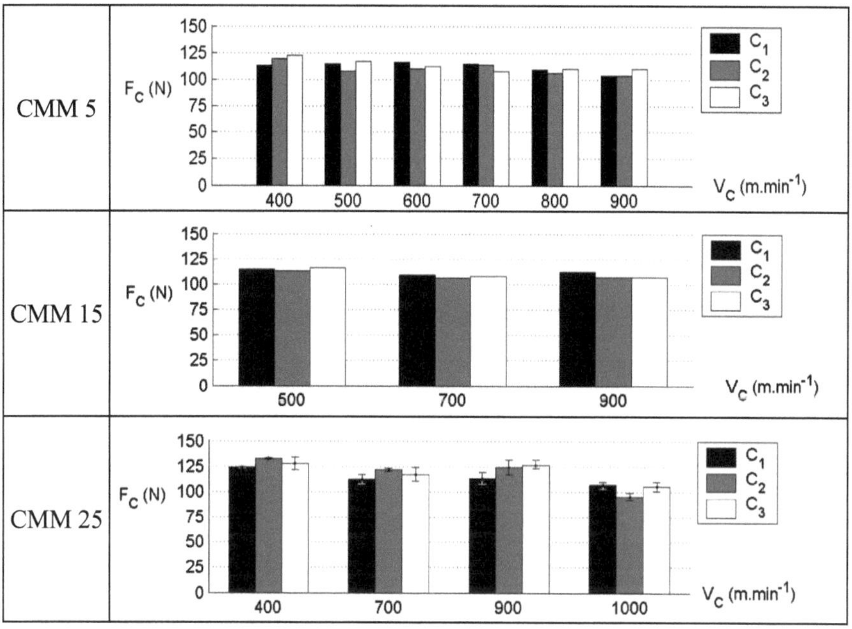

Fig. 6.5 Cutting forces for a feed rate of 0.1 mm rev^{-1}

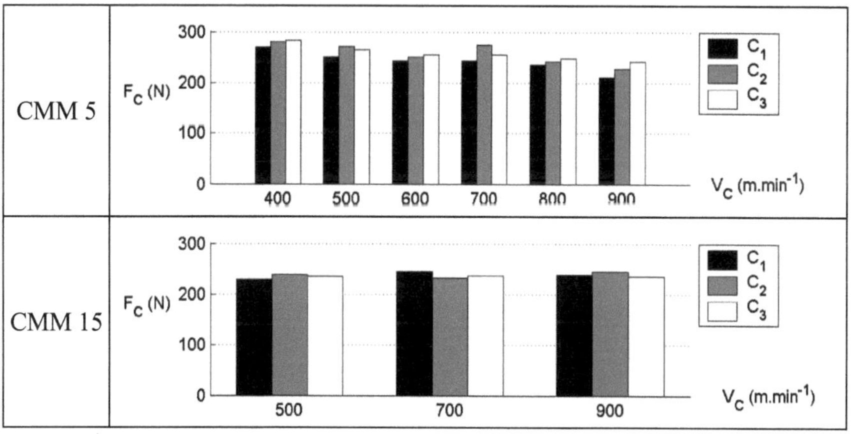

Fig. 6.6 Cutting forces for a feed rate of 0.3 mm rev^{-1}

on a reduced form on the flank face (Built-Up-Layer, BUL). This phenomenon can be explained by the temperature increasing at the tool rake face and near the cutting edge. A BUL was probably formed during the first stages of cutting, see Sanchez et al. [13] observations after dry machining aerospace aluminum alloys; then BUE takes place, however, according to the high temperature and the high pressure taking place at the tool tip, this BUE was deformed plastically and it flows

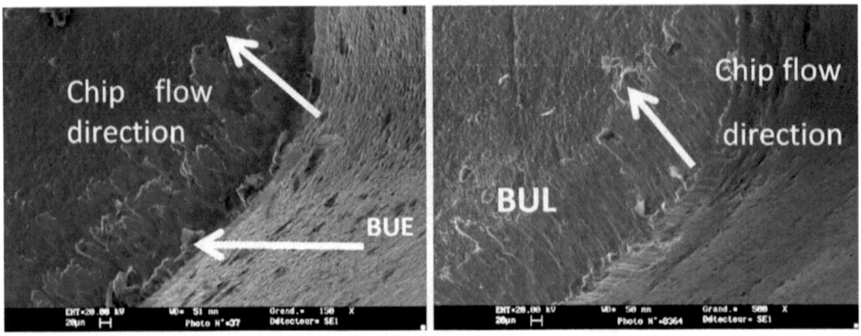

Fig. 6.7 SEM micrographs of coatings after short-duration cutting tests. **a** C2 coating, CMM 5, $V_c = 700$ m min^{-1}, $f = 0.1$ mm rev^{-1}, $a_p = 1$ mm. **b** C1 coating, CMM 15, $V_c = 500$ m min^{-1}, $f = 0.1$ mm rev^{-1}, $a_p = 1$ mm

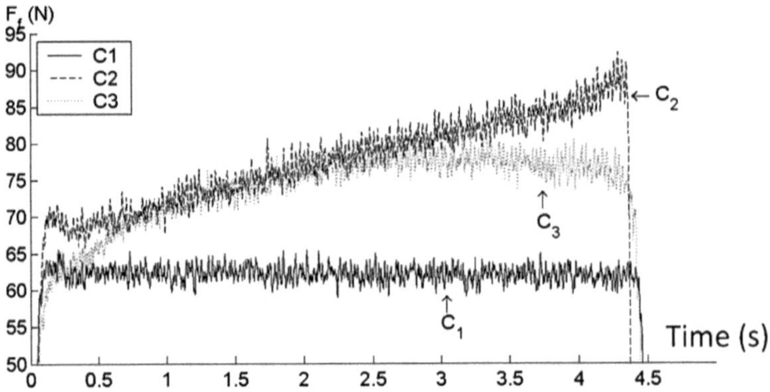

Fig. 6.8 Evolution of feed forces during machining CMM 25 (feed rate of 0.1 mm rev^{-1})

in the chip flow direction to form this multilayer deposition. Abrasive wear was not clearly depicted in the first stage of tool wear due to adhering layer which somewhat protected the tool rake face against abrasion. However, the accumulated worked material welded on tool rake face is not always stable and is repeatedly removed. This removing may produce coating peeling and then abrasive wear of the substrate material of the tool by the SiC particles.

As explained in many articles, the main wear mode in the machining of MMC is abrasive on account of the SiC particles, so the higher the level of reinforcement, the greater the wear rate. Evolution of feed forces (Fig. 6.8) on machining CMM 25 is more interesting. The feed force increases with time with the use of C2 due to tool wear evolution. The wear rate seems to be quite constant for C2, whereas for C3, the coating behavior is similar to C2 for the first half of the test, after which the feed forces remain constant. This result is most likely due to the wear of the C2-like layer at the beginning of C3 tests, followed by the lower wear rate of the

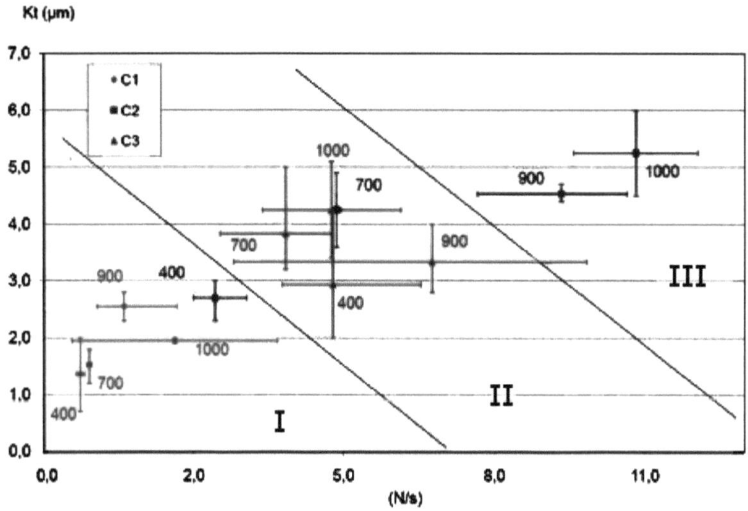

Fig. 6.9 Relationship between feed force gradient and crater wear (Kt) during machining CMM 25 for various cutting speed (value next to markers)

C1-layer. The average feed force gradient and wear were measured for each machining tests using CMM 25 (duration test was constant).

6.5 Tool Behavior and Performance

For this series of tests, the cutting speed and the feed rate were fixed. Previous section shows a little reduction of cutting forces at $V_c = 700$ m min^{-1}, so this cutting speed was selected for next tests with a feed rate of 0.1 mm rev^{-1}.

The flank wear *VB* was observed and determined regularly using an optical microscope in order to evaluate the tool life of each coated tool. The maximum allowable wear is $VB = 0.2$ mm. During these bar turning tests, spindle power consumption and feed motor current were measured continuously with WattPilot® systems. In addition, the machined surfaces were assessed qualitatively using SEM.

6.5.1 Wear Resistance Tendency

As observed from Fig. 6.8, evolution of feed forces seems to be a marker of tool wear resistance. For C1, C2 and C3 coatings, the relationship between feed force gradient and crater wear (Kt) was established in Fig. 6.9 (cutting speed was noted next to the marker). It is divided in three zones: the first one (I) corresponds to the

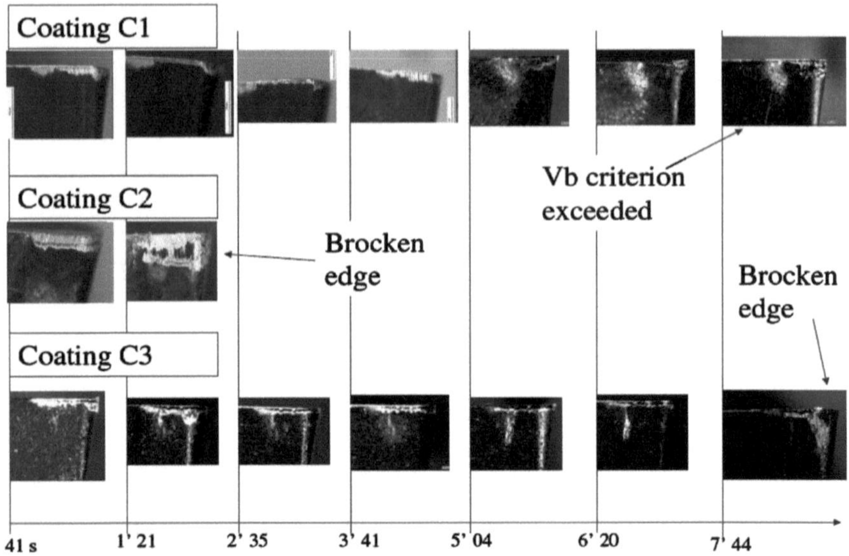

Fig. 6.10 Evolutions of the flank wear during machining CMM 15

best wear resistance, and the third zone (III) corresponds to the worst resistance. All C1 coating tests are in the first zone, all C3 are in the second and C2 tests are split, with about half falling in the third zone. This figure permits to envisage results of longer tests.

6.5.2 Tool Life Influence of Nano-Structured Coatings

For coated tools C1, C2 and C3, the tool wear was regularly evaluated by optical microscope observations as shown in Fig. 6.10. As shown previously (Fig. 6.9), the C2-coating gave the worst result. A complete failure of the tool was observed after 80 s. For the other coated tools, the tool life limit was attained for a cutting duration of about 7 min, after an abrasive wear process for C1-coating and after coating peeling and cutting edge attrition for C3-coating (Fig. 6.11).

Figure 6.12 presents the evolution of power spindle consumption during machining CMM 5 and 25. A device was used to measure power consumption for CMM 5 and another one for CMM 25 machining tests, it explains the difference in measured values but the evolution is still significant. Evolution of cutting power for coatings C2 and C3 did not change with the level of reinforcement. With C2, the spindle power requirement increases continuously with time and independent with respect to the level of reinforcement. During machining CMM 5, power consumption is constant for C3 and C1, whereas it increases during the C1 test on CMM 25.

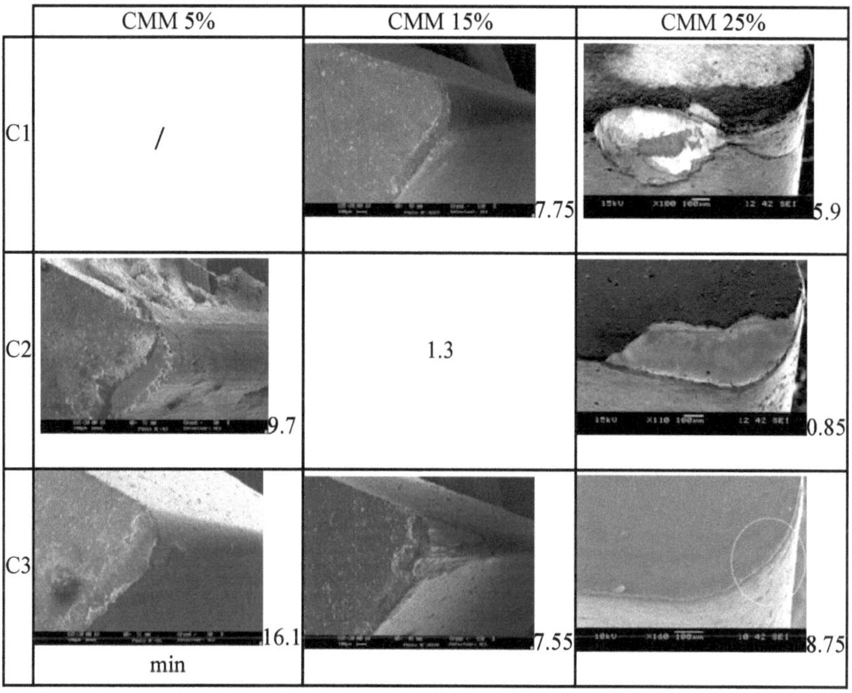

Fig. 6.11 SEM micrographs of C1, C2 and C3 coated tools (indication of *cutting times*)

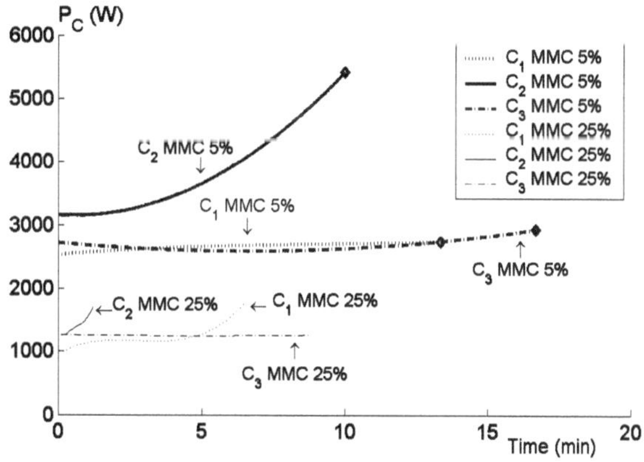

Fig. 6.12 Evolution of spindle power consumption (CMM 5 and CMM 25 for CVD tools C1, C2 and C3)

Observation of the edge reveals high-wear rate for C2 and C1, and breakage of coating C3. The increase in the reinforcement level leads to frequent coating peeling (Table 6.2 and Fig. 6.11). The wear rate of the edge leads to a constant decrease in sharpness, which explains the increase in the power required during cutting process using the C2 coating. C2 Tests were stopped due to a lot of work piece material adhering on flank face.

Table 6.2 Tool life and wear or damage mode for C1, C2 and C3 coating tools

	CMM 5%	CMM 15%	CMM 25%
C1	/	7.75 (flank wear)	5.9 (coating peeling + notch wear)
C2	9.7 (flank wear)	1.3 (flank wear)	0.85 (insert breakage)
C3	16.1 (edge breakage)	7.55 (edge breakage)	8.75 (coating peeling)

For all tests, the coating C2 presents the worst results with an important adhesive layer on the flank face. This adhesion phenomenon exists also on the rake face, creating a BUL and a BUE [13] which was taken off regularly by chip flow with abrasives particles. The smooth surface of C2 induced a great adhesion area, which put significant stress on the diamond film, resulting in the striping of diamond particles from tool tip.

C3 never attained the flank wear criterion, but had a notch wear when machining CMMC 15 at the same location as compared to machining CMM 5. This place corresponds to the highest temperature area. The increase in temperature should decrease the stickiness of the two diamond layers, resulting in peeling of the film.

Bogli [16] have shown the lack of representativeness of coating roughness and the importance of its morphology. Important angles of asperities generate a large interfacial layer which reduces the coefficient of friction. This explains the best results of C1 compared to C2, this difference was increased by an adhesive wear on C2.

During self-mated sliding tests, Schade [17] studied cauliflower structure and fine-grained diamond with small faceted crystallites. Cauliflower structure presents an important coefficient of friction due to a large contact area (contact cauliflower–cauliflower) with a low-wear rate, unlike fine-grained coating. These articles show the importance of the morphology of contact at the grain scale and the scale of asperities. C3 coating merges a smooth contact (C2-like layer) with a structure with asperities due to the sub layer C1-like. The macro-structure generates a protective layer without adhesion wear, the smooth contact deflects particle and reduces associated wear.

6.5.3 Tool Life Test

Figure 6.13b presents tool life of coated tools while machining MMC with different levels of reinforcement. For CMM 35, C1 and C3, coatings had not reached the wear limit after 25% time more than C2 and C1810. As expected the wear rate is proportional to the percentage of reinforcement as it can be observed from Figs. 6.13b and 6.14. The abrasive action of SiC particles can be explained easily from the SEM micrographs of tool wear. In addition, this micrograph shows the adhesion problem of this CVD tool. The Fig. 6.13 gives an estimate of performance of each coating. According to Fig. 6.13b, C3 seems to have higher-wear resistance than other coatings.

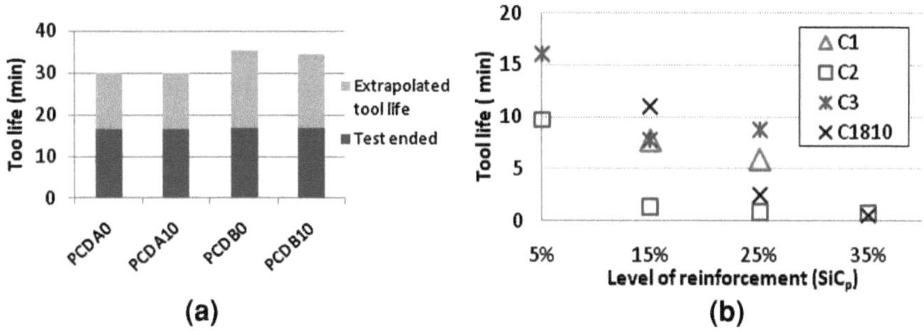

Fig. 6.13 Tool life while machining CMM 25 for PCD tool (**a**), and for all composites with CVD tool (**b**)

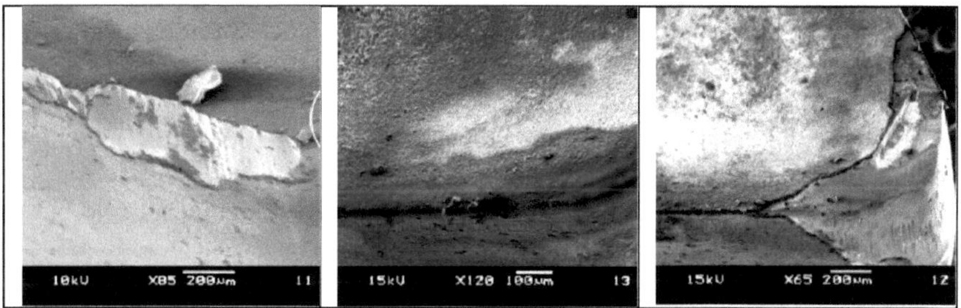

Fig. 6.14 SEM micrographs of C1810 coated tool after tool life test (indication of *cutting times*)

Table 6.3 Flank wears *VB* after about 17 min of machining CMM 25

	$\gamma = 0°$ (mm)	$\gamma = 10°$ (mm)
PCD A	0.11	0.11
PCD B	0.095	0.098

Better resistance of PCD tools to abrasive wear was already known, but it is confirmed from Fig. 6.13. In spite of machining of CMM 25 for more than 15 min, the wear limit was not reached. An approximate calculation can be realized with few hypotheses: flank wear evolves linearly with cutting times and PCD tool was worn as soon as $VB = 0.2$ mm. According to massive structure of PCD insert (thickness of diamond particles sintered together is about 1 mm, compared to coating of 6 μm), those hypotheses seem realistic.

The flank wear was measured with a Wyko® NT3300 Optical Profiler, using the white light interferometry principle [18], and results are given in Table 6.3.

Compared to coated tools, tool life of PCD tools are three times longer during machining MMC 25%. In addition, because of the massive structure of PCD tools unlike CVD tools, tool life criterion $VB = 0.2$ mm could be restrictive for PCD tool. Criterion should be adapted to integrity of machined surface (roughness, residual stresses, etc.).

6.6 Environmental Impact of Machining

Machining process generates fine, ultra fine and nano particles. Inhalation of polluted air settles particles along airways in relationship to particle size. In order to limit health diseases for worker, many countries increase legislation on air quality at workplace. The World Health Organization (WHO) is the directing and coordinating authority for health within the United Nations system. It is responsible for providing leadership on global health matters. This organization has published a study concerning airborne dust in 1999 [19] and during its Sixtieth World Health Assembly [20], recommendations were made to promote implementation of the "global plan of action on workers' health 2008–2017" at national and international levels.

Songmene et al. [21, 22] estimate that friction plays a significant role in dust generation and that there is several sources such as primary shear zone [23, 24], chip surface and tool–work piece interface. In this section many dust sources have been evaluated:

- Friction at shear plane;
- The influence of the tribological conditions at tool–chip interface;
- Influence of tool wear resistance.

Dust emission quantification was made during machining. An aerosol spectrometer was used to detect airborne aerosol particles in the size range of 0.3–20 μm in 15 size channels in real time and to represent the results in particle mass. To compare the results, an index was developed by Khettabi et al. [24] called Dust unit (Du). Du is the ratio of the dust mass to the mass of chip removed from the workpiece material.

6.6.1 Dust Generation at Shear Plane

Cutting parameters were fixed at a cutting speed of 700 m min^{-1}, feed rate of 0.1 mm rev^{-1} and depth of cut of 1 mm, except for the C3 coating where depth of cut was about 0.2 mm.

Figure 6.15 presents Du for particle sizes below 4 μm calculated for the machining of MMC 25%. This figure shows that machining with C2 produces 45% more dust than C1. Ballout et al. [25] established a correlation between chip

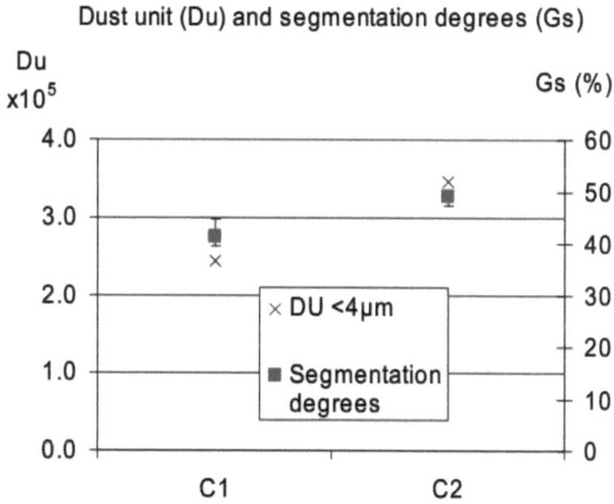

Fig. 6.15 Du (particles <4 μm) and segmentation degrees G_S ($V_c = 700$ m min^{-1}, $f = 0.1$ mm rev^{-1} and $a_p = 1$ mm)

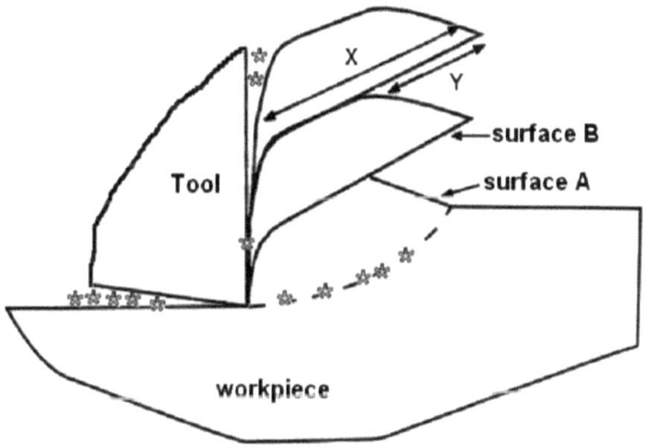

Fig. 6.16 Localization of dust generation (*stars*)

formation and dust emissions. They explained that dust production during machining is due to the formation of micro-bands within the chip. The different dust emission localizations described by Ballout et al. [25] are summarized in Fig. 6.16 and symbolized by stars.

The analysis of the chip was made to explain different dust emission measurements between the two coatings. Figure 6.16 represents a schematic of the saw-toothed chip encountered during those tests. Surface A corresponds to an undeformed surface, whereas surface B is a part of the catastrophically failed surface separated from the following segment due to intense shear. This last surface is one of the places where dust emissions occur [25]. Chip surface observations are shown in Fig. 6.17. The surface A are obtained with coated tools and are defined by planes. The surface B is really different for each coating. For coating C1 test, surface B seems to be a plane so the surface area of shear "plane" corresponds to chip section. For C2 and C3 tests, Fig. 6.17 reveals that surface B is wavy. The consequence is that the shear "plane" areas are larger than with C1 coating test, so the sliding interface area that produces dust is greater and it results

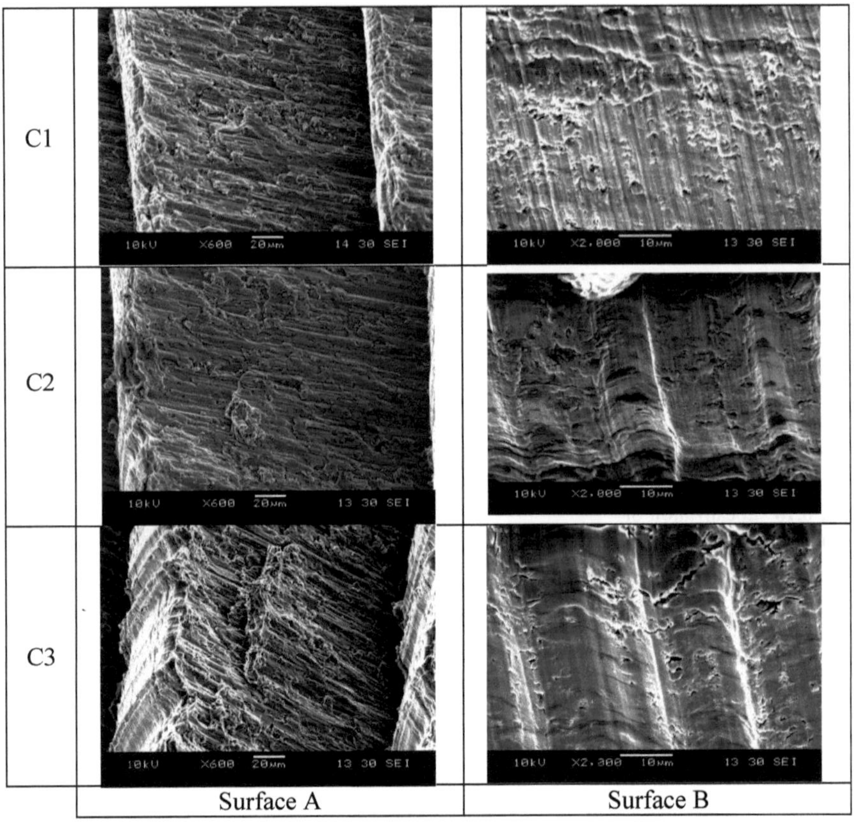

Fig. 6.17 SEM micrographs of chips (CMM 25, $V_c = 700$ m min^{-1}, $f = 0.1$ mm rev^{-1} and $a_p = 1$ mm)

in an increase of dust emission. Balout et al. [25] took into account only the density of the shear plane because the shear planes in their study present similar morphology. In this study the density is constant between all tests, but the shear plane area is affected by the coating structure. Surfaces B of C2 and C3 tests are corrugated, while for C1 test this surface is almost flat.

Schulz et al. [26] used a parameter called chip segmentation degrees (Gs) to measure the impact of processing on saw-toothed chips. This parameter is defined as Gs = Y/X with X and Y represented in Fig. 6.16. The increase in the segmentation degrees (Fig. 6.15) indicates that Y is longer as compared to X (please check) for the same chip height. A part of dust comes from shear plane location, an increase of Y value means that dust is less confined and is easily disseminated.

For C2 test, the wavy shear plane increases the sliding contact area and consequently the dust emission over the C1 test. Additionally the higher chip segmentation degree facilitates the spread of dust.

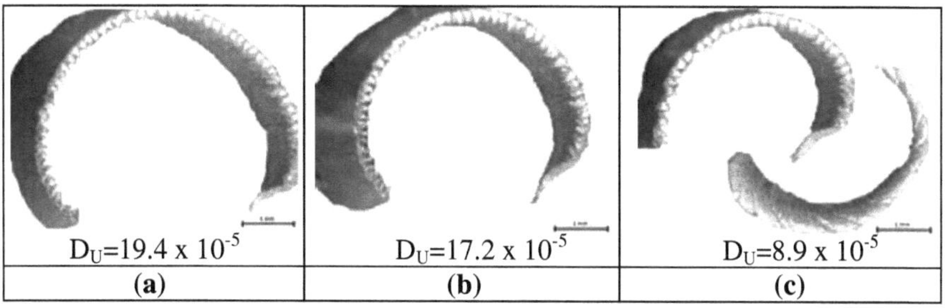

Fig. 6.18 Chip morphology and Du associated to machining CMM 15 with PCD A_{10} after a cutting time of 2 min (**a**), 3 min (**b**) and 9 min (**c**)

6.6.2 Tribological Conditions at Tool Chip Interface

In this section, cutting parameters were fixed at a cutting speed of 700 m min^{-1}, feed rate of 0.1 mm rev^{-1} and depth of cut of 1 mm for PCD and CVD tools.

6.6.2.1 Chip Morphology and Dust Emission

For each machining test, chips was kept and observed on optical microscopy. Figure 6.18 presents chips obtained during machining MMC 15% with PCD tools for different cutting time and the corresponding Du. The scale is similar for a, b and c and represents 1 mm. In this figure the radius of curvature seems to evolve in the same way as the amount of dust generated. During machining the chip with the smallest radius (Fig. 6.18c) generated half the dust than the chip with largest radius (Fig. 6.18a). This difference corresponds also due to different cutting time. More dust was generated at the beginning of the machining.

In order to confirm this impression, all chips obtained during machining MMC 15% with PCD were measured and represented in a graph with the value of Du corresponding to the test. Figure 6.19 aims to show the relationship between chip radius (abscissa) and Du (ordinate). The regression lines have been added to highlight the link between these two parameters. This figure put in evidence this trend, but we must keep in mind that the measurement of radius of curvature presents measurement uncertainties, that can be explained by the gap between regression lines and some measurement points.

Childs [27] describes the curly chip formation with the slip-line field method using the previous study of Kudo [28]. Childs explains that tool–chip interface had important influence on contact length on rake face, chip thickness and radius of curvature. The curvature of the chip provides from the existence of a compression area between shear plane and tool chip interface, and a more rigid zone at the free surface of the chip. The different behavior of these two areas causes the curvature of the chip. In more recent studies, Dewhurst [29] explained that the contact shear stress along the tool rake face is an important parameter. Increasing friction along

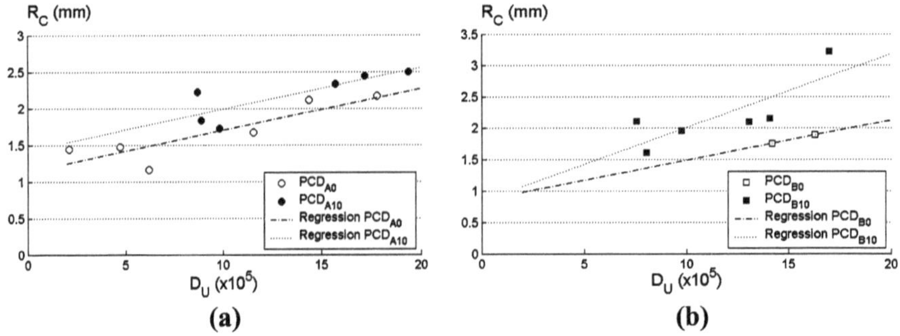

Fig. 6.19 Chip radius and Du measured for PCD A (**a**) and PCD B (**b**) (rake face of 0 and 10°) during machining MMC 15%

the tool rake face reduces the chip curvature. So the higher the friction stress, the greater is the chip radius, and the dust production increases. The evolution of DU observed in Fig. 6.18 was linked to the contact at tool–chip interface. At the beginning (Fig. 6.18a) the rake face is free of work piece material, but adhesion of matrix material generates BUE and BUL. Once BUE and BUL installed, contact conditions are modified and friction at the interface decreases as dust emission.

Friction at the interface added to contact length influence dust emission. Figure 6.20a and c shows rake face. In those figures contact length was measured between adhesive material and tool edge. During machining of CMM 15 it was equal to 0.2 and 0.175 mm for machining of CMM 35. The lower-contact length can be explained by the important chip segmentation of CMM 35. The segmentation mechanism on machining MMC is mainly controlled by cracks propagation from a particle to another one. The higher-chip segmentation was observed in Fig. 6.20b and d, consistently to level of reinforcement. This causes a reduction of D_U.

6.6.2.2 Influence of Edge Acuity

As described previously, the influence of interface is important and the quality of sliding contact at the interface evolves with time (Fig. 6.18a and c). Longer tests were conducted all along tool life of CVD tool during machining CMM 15 and CMM 25. Spindle power consumption and dust emission were monitored continuously and evolutions were presented in Fig. 6.21.

The dust level at the beginning of machining with CVD tool is lower in relation to the amount produced with PCD (Fig. 6.21) at the same time. This difference is due to the important difference of surface morphology of rake face (difference of C1810 profile at Fig. 6.3 and profile of PCD A at Fig. 6.4). PCD tools have a high flatness with almost no peaks or valleys (total height of the roughness profile $R_t = 0.6$ μm) whereas on rake face of CVD tool, diamond grains can be observed ($R_t = 2.1$ μm). It mainly comes from variations in manufacturing processes

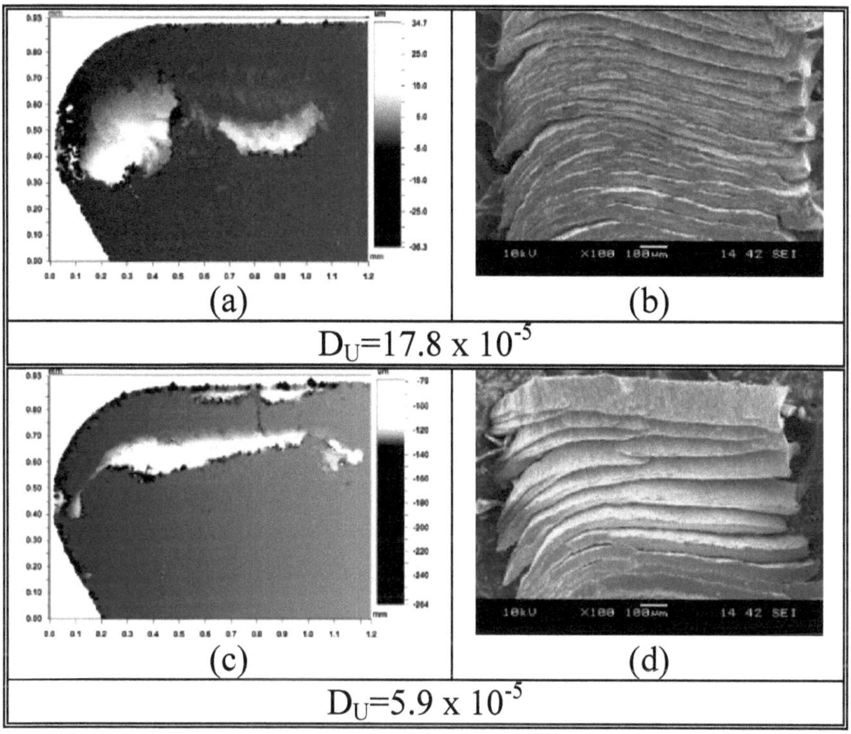

Fig. 6.20 PCD A_0 after machining CMM 15 (**a**), CMM 35 (**c**), segmented chip of CMM 15 (**b**) and CMM 35 (**d**)

Fig. 6.21 Evolution of D_U during machining CMM 25 for C1810 and PCD A10

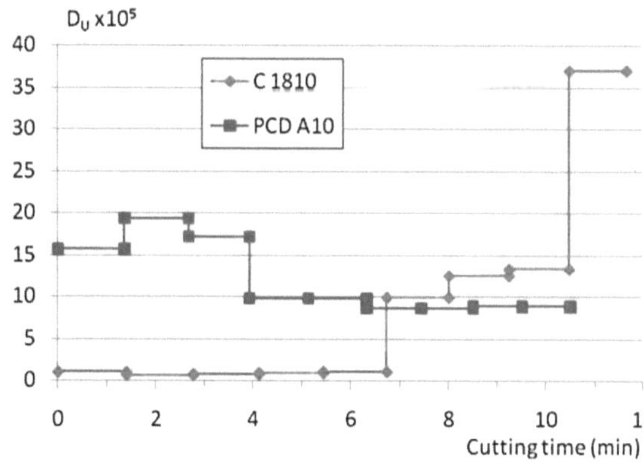

PCD/CVD. The high flatness of PCD tools increases contact area, so friction increases too and more dust is generated than with CVD tool.

Tests were stopped because of tool failure (tool breakage after 11 min on CMM 15, coating peel off after 2.5 min on CMMC 25, micrographs of tools are shown in Fig. 6.14). Evolution of Du is similar to power consumption evolution on both

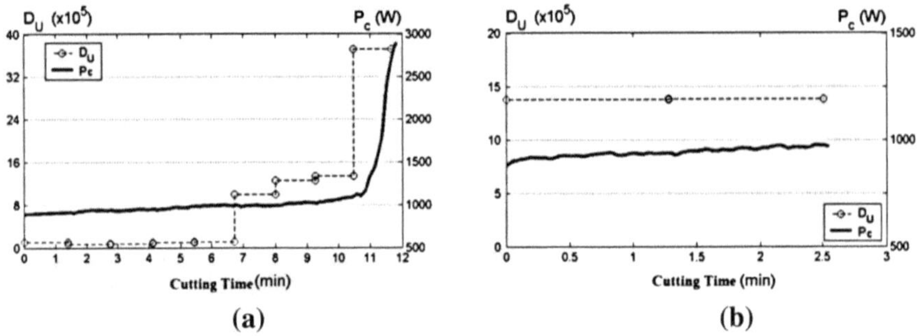

Fig. 6.22 Evolution of Du (D_U) and spindle power consumption (Pc) during machining CMM 15 (**a**), and CMM 25 (**b**) with CVD tool C1810

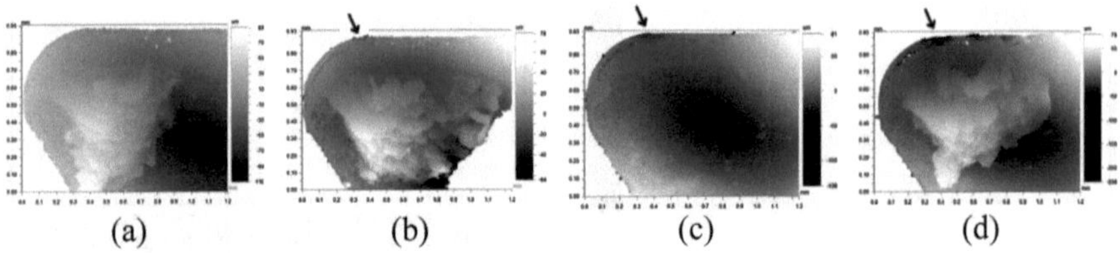

Fig. 6.23 Rake face of CVD tool (1810) after machining MMC 15% during 6.8 (**a**), 8 (**b**), 9.2 (**c**) and 10.5 min (**d**)

graph of Fig. 6.22. However, Du increases continuously after 6.8 min whereas power consumption is constant during the first 11 min. To explain this difference, edge and tool rake face were observed with an optical interferometer (Fig. 6.23).

In Fig. 6.23a, some material adheres on rake face but tool edge is intact. In others parts of this figure (b, c and d) damage appears at the junction of the corner radius and the linear edge. It is indicated with arrows on each observation. Damaged area expands, in the same way as dust generation. Dust emission seems to be more sensitive to wear edge than consumed spindle power. During machining CMM 25, there is no evolution on dust measurement and on power consumption, it could be explained by apparition of decohesion of coating at the end of test. It seems to occur when tool is going out of workpiece material.

6.6.3 *Influence of Tool Wear Resistance*

Figure 6.24 presents the evolution of Du compared to level of reinforcement of the matrix. For CVD tools, an increase in the level of reinforcement leads to a larger Du. This evolution can be explained by the higher-wear rate induced by larger number of particles. Songmene et al. [21, 22] have similar results in drilling process with sharp drill and worn drill. A worn edge induces more important friction phenomena which generate more dust as compared to a sharp drill. It is not just the total quantity of dust that evolves with the level of reinforcement, aerosol metrology also changes.

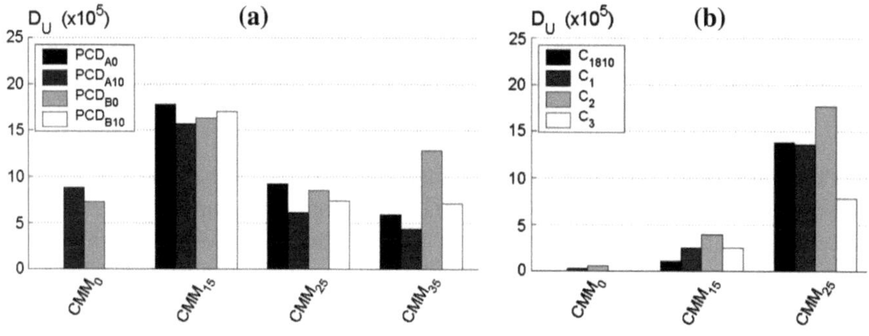

Fig. 6.24 Du for different levels of reinforcement and different tools: PCD (**a**), and CVD (**b**)

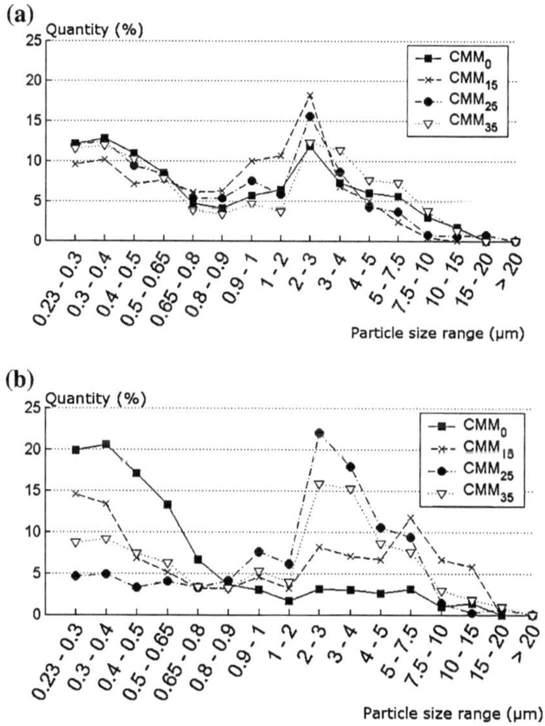

Fig. 6.25 Size distribution of particles emitted during the cutting process with PCD B0 (**a**) and CVD C2 (**b**)

Figure 6.24 presents repartition of dust particles generated during machining MMC with CVD tool C2. Two groups of particles are mainly present: small particles (from 0.23 to 0.8 μm diameter) and medium particles (from 2 to 7.5 μm diameter). The proportion of medium particles increases, whereas proportion of small particles decrease with the level of reinforcement. The size range of medium particles is similar to the size range of diamond particles (Fig. 6.3) and particles of SiC (Fig. 6.2). This group of medium particles seems to be composed of diamond grains taken off from coating (wear), and of SiC particles extracted from bulk material by tool action. The extraction particles process was explained by Li et al. [30]. This phenomenon grows with the proportion of SiC particles.

Physical presence of diamond grains in dust emission can be linked to the wear rate of CVD tools. Figure 6.13b gives an estimate of performance of each coating. According to this figure, C3 seems to have higher-wear resistance than other coatings. This property associates with the lowest-dust emission value for CVD tools (Fig. 6.24b) is in agreement with explanation of evolution of Du.

The metrology of dust generated with PCD tools is different than with CVD tools. Figure 6.25a reveals only modification for particles larger than 1 μm during machining with PCD tool. Whereas with CVD tools the percentage of particles larger than 3 μm increases with the quantity of SiC particles, and proportion of particles in the size range 2–3 μm decreases. Those larger particles correspond to SiCp extracted from aluminum matrix during cutting process. Extraction localized at the tool edge was previously explained (preceding paragraph and [30]) but according to Arumugam et al. [31] chip segmentation generates also particles emission.

6.7 Conclusion

Because of abrasive action of SiC particles, diamond tools are the most suitable tools to machine MMC.

The performances of CVD tools are varying with nano-structure of the film and the ability of adhesion of the film. PCD tools have more constant results, and their higher cost seems to be offset by significant longer-tool life.

Friction seems to be the main generator of dust during machining process. It takes place at several locations.

- *At the shear plane*. The smooth coating emits more dust than the rough one because shear planes are wavier. In addition, higher segmentation degrees of the chip produced with the tool with a smooth coating facilitate the spread of dust.
- *At tool chip interface*. The variation of curvature chip revealed the modification of friction stress on tool rake face. Larger the radius of curvature is, the greater the friction stress is, the more there is dust produced.
- *On rake face*. Its morphology has an important impact on dust generation, and it generally depends on manufacturing process. PCD tools present high flatness on tool face, which leads to a "running in" period with important emission of dust, contrary to lower dust generation of CVD tool at the beginning of machining, because of a more irregular surface.
- *At tool edge*. Similarly to cutting power, dust emitted increases with wear. However, the Du Index is more sensitive to edge damage than power consumption and detects little modification on cutting process.

Properties and composition of the workpiece material can also induce dust generation. The crack propagation properties and the abrasive action of SiC particles are important:

- The high-wear rate of MMC upon diamond coating tools produces dust with particles of SiC extracted and diamond grains detached. The better the coating resistance is, the lower the quantity of dust.
- The good resistance to abrasive wear of PCD tools increases the importance of tool–chip interface on dust emission. Friction at contact area chip/rake face seems to be the dominant dust generator with PCD tools.
- The percentage of reinforcement has various effects on dust emission depending on the tool type. The main parameter for CVD tools is the wear rate: quantity of dust emission increases with the level of reinforcement. On the other hand PCD tools are more sensitive to tool–chip interface. The increase of reinforcement rate produces a more segmented chip which reduced the contact length and decreases dust emission.

References

1. Manna A, Bhattacharayya B (2003) A study on machinability of Al/SiC metal-matrix composites. J Mater Process Technol 140:711–716
2. Chambers AR (1996) The machinability of light alloy MMCs. Composites Part A 27A: 143–147
3. El-Gallab M, Sklad M (1998) Machining of Al/SiC particulate metal-matrix composites. Part I: Tool performance. J Mater Process Technol 83:151–158
4. El-Gallab M, Sklad M (1998) Machining of Al SiC particulate metal matrix composites. Part II: Work piece surface integrity. J Mater Process Technol 83:277–285
5. Iullanoa L, Settineria L, Gattob A (1998) High speed turning experiments on metal matrix composites. Composites Part A 29A:1501–1509
6. Andrewes JEC, Feng Hsi-Yung, Lau WM (2000) Machining of an aluminium/SiC composite using diamond inserts. J Mater Process Technol 102:25–29
7. Cappelli E, Pinzari F, Ascarelli P, Righini G (1996) Diamond nucleation and growth on different cutting tool materials: influence of substrate pre-treatments. Diamond Relat Mater 5:292–296
8. Belmonte M, Ferro P, Fernandes AJS, Costa FM, Sacramento F, Silva RF (2003) Wear resistant CVD diamond tools for turning of sintered hard metals. Diamond Relat Mater 12:738–743
9. Miranzo P, Osendi MI, Garcia E, Fernandes AJS, Silva VA, Costa FM, Silva RF (2002) Thermal conductivity enhancement in cutting tools by chemical vapor deposition diamond coating. Diamond Relat Mater 11:703–707
10. Durante S, Rutteli G, Rabezzana F (1997) Aluminium based MMC machining with diamond coated cutting tools. Surf Coat Technol 94–95:632–640
11. Vandevelde TCS, Vandierendonck K, Van Stappen M, Du Mong W, Perremans P (1999) Cutting applications of DLC, hard carbon and diamond films. Surf Coat Technol 113(1–2): 80–85
12. Shen CH (1996) The importance of diamond coated tools for agile manufacturing and dry machining. Surf Coat Technol 86–87(2):672–677

13. Sanchez JM, Rubio E, Alvarez M, Sebastian MA, Marcos M (2005) Microstructural characterization of material adhered over cutting tool in the dry machining of aerospace aluminum alloys. J Mater Process Technol 164–165:911–918
14. Hung NP, Yeo SH, Oon BE (1997) Effect of cutting fluid on the machinability of metal matrix composites. J Mater Process Technol 67:157–161
15. Sutherland JW, Kulur VN, King NC (2000) An experimental investigation of air quality in wet and dry turning. Ann CIRP 49(7):61–64
16. Bogli U, Blatter A, Pimenov SM, Obraztsova ED, Smolin AA, Maillat M, Leijala A, Burger J, Hintermann HE, Loubnin EN (1995) Diamond Relat Mater 4:1009
17. Schade A, Rosiwal SM, Singer RF (2007) Influence of surface topography of HF-CVD diamond films on self-mated planar sliding contacts in dry environments. Surf Coat Technol 201:6197–6205
18. Devillez A, Lesko S, Mozer W (2004) Cutting tool crater wear measurement with white light interferometry. Wear 256(1–2):56–65
19. World Health Organization (WHO) (1999) Hazard prevention and control in the work environment: airborne dust. WHO/SDE/OEH/99.14
20. World Health Organization (WHO) (2007) Workers' health: global plan of action, sixtieth world health assembly. WHA60.26
21. Songmene V, Balout B, Masounave J (2008) Clean machining: experimental investigation on dust formation. Part I: Influence of machining parameters and chip formation. Int J Environ Conscious Des Manuf 14(1):1–16
22. Songmene V, Balout B, Masounave J (2008) Clean machining: experimental investigation on dust formation. Part II: Influence of machining strategies and drill condition. Int J Environ Conscious Des Manuf 14(1):17–33
23. Kremer A, El Mansori M (2009) Influence of nanostructured CVD diamond coatings on dust emission and machinability of SiC particle-reinforced metal matrix composite. Surf Coat Technol 204:1051–1055
24. Khettabi R, Songmene V, Masounave J (2007) Effect of tool lead angle and chip formation mode on dust emission in dry cutting. J Mater Process Technol 194:100–109
25. Balout B, Songmene V, Masounave J (2007) An experimental study of dust generation during dry drilling of pre-cooled and pre-heated workpiece material. J Manuf Processes 9(1):23–34
26. Schulz H, Abele E, Sahm A, Institute of Production Engineering and Machine Tools, University of Technology, Darmstadt, Germany
27. Childs THC (1971) A new visio-plasticity technique and a study of curly chip formation. Int J Mech Sci 13:373–387
28. Kudo Hideaki (1965) Some new slip-line solutions for two-dimensional steady-state machining. Int J Mech Sci 7:43–55
29. Dewhurst P (1978) On the non-uniqueness of the machining process. Proc R Soc London, Ser A 360:587–610
30. Xiaoping L, Seah WKH (2001) Tool wear acceleration in relation to workpiece reinforcement percentage in cutting of metal matrix composites. Wear 247:161–171
31. Arumugam P, Malshe A, Batzer S (2006) Dry machining of aluminum silicon alloy using polished CVD diamond coated cutting tools inserts. Surf Coat Technol 200:3399–3403

Chapter 7
Computational Methods and Optimization in Machining of Metal Matrix Composites

V. N. Gaitonde, S. R. Karnik and J. Paulo Davim

This chapter deals with the importance of mathematical modeling and need for optimizing the process. Further, case studies involving the various modeling and optimization techniques applied to machining of metal matrix composites are also discussed.

7.1 Introduction

The main objective of any machining process is to obtain the desired level of quality characteristic for the finished component. In order to study the machining behavior, it is essential to establish the relationship between the controlled parameters and the required quality characteristic. On the other hand, to optimize the machining process, the proper setting of control parameters (factors) on the performance measure (response) is necessary. Both modeling and optimization

V. N. Gaitonde (✉)
Department of Industrial and Production Engineering,
B. V. B. College of Engineering and Technology,
Hubli, 580031, Karnataka, India
e-mail: gaitondevn@yahoo.co.in

S. R. Karnik
Department of Electrical and Electronics Engineering,
B. V. B. College of Engineering and Technology,
Hubli, 580031, Karnataka, India
e-mail: karniksr@yahoo.com

J. Paulo Davim
Department of Mechanical Engineering, University of Aveiro,
Campus Santiago, 3810-193 Aveiro, Portugal
e-mail: pdavim@ua.pt

J. P. Davim (ed.), *Machining of Metal Matrix Composites*,
DOI: 10.1007/978-0-85729-938-3_7,

require efficient planning of experiments for minimizing the number of experiments, which in turn reduces the time and cost involved in the experimentation. The design of experiments (DOE) is extensively used in the planning of experiments involving several factors, where it is necessary to investigate the joint effect of the factors on a response variable [1, 2]. The DOE takes levels of several input parameters to formulate the different combinations at which the output parameters are to be observed or computed. There are several types of DOE available, which are based on statistical theory [1, 2]. The selection of proper DOE basically depends on the purpose of experimentation, which includes the development of modeling or/and optimization of machining process.

7.1.1 Importance of Mathematical Modeling

In any machining process, a mathematical model is constructed for prediction, optimization and controlling the process. A regression model is built to a set of sample data, which describes the relationship between the quality characteristic (y) and input control factors (x_1, x_2, ..., x_k). The true functional relationship is given by:

$$y = \varphi(x_1, x_2, x_3, \ldots x_k) \tag{7.1}$$

In most of the cases, true functional relationship is not known and hence an appropriate function to approximate φ is to be chosen. Normally, low order polynomial models such as first and second orders are largely used as approximating functions [1, 2] and these empirical regression equations are called as response surface models.

Identifying and fitting an appropriate response surface model from experimental data requires some knowledge of DOE, regression modeling techniques and elementary optimization methods [1, 2]. The response surface methodology (RSM) integrates all of the above. These empirical models serve to provide information about the properties of the system from which the data are taken [1, 2]. There are many other situations in which a mathematical model is used for process optimization.

There are many types of experimental designs available, which are mainly based on number of factors and their levels. The appropriate DOE to be selected is based on three criteria, namely, purpose (model development/optimization), nature of model (linear or nonlinear model) and the number of factors and the levels defined for each of the factors [3]. The DOE can be classified as full-factorial design, fractional-factorial design, orthogonal array, central-composite design and Box–Behnken design [1–4].

7.1.2 Need for Optimization

The optimization of a system (product/process) design means determining the best architecture, the best parameter values and the best tolerances [4]. However, the optimization of any process still remains one of the most challenging problems because of its high complexity and non-linearity while solving it and most of the engineering design problems are multi-objective in nature. Hence, designing the system at the required quality characteristic is an economical and technological challenge to the engineer or scientist.

The conventional methods of optimization do not fare well over a broad spectrum of problem domains and also not efficient when the practical search space is too large. Moreover, these methods are also not robust and tend to obtain a local optimum solution. In order to meet the above requirement, a systematic and efficient method of design optimization for performance, quality and cost is crucial [4]. The techniques for solving the optimization problems can be categorized as:

- Taguchi robust design optimization technique i.e., by allowing the process optimization using Taguchi technique with minimum number of experiments without the need for the development of models [4].
- Model-based optimization techniques i.e., by using the empirical models to obtain the predictions of the response of interest and then to find the best compromises among the multi-objectives for the system through the non-traditional optimization tools like genetic algorithms (GA), simulated annealing (SA), ant colony optimization (ACO) and particle swarm optimization (PSO) [5].

7.2 Modeling Approaches

The metamodels are the empirical expressions, which are used to obtain the relationships relating the control factors to the performance characteristics of interest. The data obtained from the statistical experimental design is utilized to construct the models. The advantage of the metamodel is that once it is created, it can yield large amounts of predictions. The most commonly used metamodels are response surface model, fuzzy logic, artificial neural networks and neuro-fuzzy inference systems. However, the models based on RSM and artificial neural network (ANN) has recently gained much attention in studying and analyzing the behavior of machining processes.

7.2.1 Modeling Based on Response Surface Methodology

The RSM using DOE proved to be an efficient mathematical modeling tool [1, 2]. The methodology not only reduces the cost and time, but also gives the required information about the main and interaction effects of the factors with reduced number of experiments. The RSM is a collection of mathematical and statistical techniques, which are useful for building the mathematical models and analyzing the problems that provide an overall perspective of the system response within the design space [1, 2]. The mathematical model of the quality characteristic to the control factors can be predicted by employing the multiple regression analysis with minimum number of experiments planned through DOE. The RSM refers not merely to the use of a response surface as a multivariate function but also to the process for determining the polynomial coefficients.

The construction of RSM models involves three main steps, namely, choosing a functional form for the model representation, fitting the model to the observed data and the statistical validation for the response surface [3].

After conducting a set of experiments to obtain the quality characteristics or outputs according to the experimental designs, the next step is to take the vectors of input control factors (x) and the corresponding responses (y) for fitting the appropriate model. Typical response surface model limits the order of polynomial to one or two since the low-degree models contain fewer terms than the higher-degree models and thus, require fewer experiments to be performed [1–3].

The first-order model is likely to appropriate when the engineer or scientist is interested in approximating the true response surface over relatively small region of independent variable space in a location where there is little curvature in true response function [1, 2]. The first-order model is given by [1, 2]:

$$y = \beta_0 + \sum_{i=1}^{k} \beta_i x_i \tag{7.2}$$

where, βs' are regression coefficients and k is the number of input variables. If there is curvature in the system i.e., the response exhibits nonlinear behavior, the second-order polynomial is considered as a response surface model and is given by [1, 2]:

$$y = \beta_0 + \sum_{i=1}^{k} \beta_i x_i + \sum_{i=1}^{k} \beta_{ii} x_i^2 + \sum\sum_{i<j} \beta_{ij} x_i x_j \tag{7.3}$$

The second-order model can take on a wide variety of functional forms, so it will often work well as an approximation to the true response surface.

The values of regression coefficients of linear, quadratic and interaction terms of the model are determined by [1, 2]:

$$b = \left(X^{\mathrm{T}} X\right)^{-1} X^{\mathrm{T}} Y \tag{7.4}$$

where, b: regression coefficient, X: calculation matrix that contains main and interaction terms, X^{T}: transpose of X, $\left(X^{\mathrm{T}} X\right)^{-1}$inverse matrix of a $\left(X^{\mathrm{T}} X\right)$ and Y: matrix of measured response.

The analysis of variance (ANOVA) [1, 2] technique is used to check the adequacy of the developed model for the desired confidence interval. The ANOVA table includes the sum of squares (*SS*), the degrees of freedom (*DF*) and the mean square (*MS*). In ANOVA, the contribution for *SS* is from the first order terms, the second-order terms and the residual error. The *MS* are obtained by dividing the *SS*

Table 7.1 Process parameters and their levels

Parameters	Unit	Levels			
		1	2	3	4
Cutting speed (v)	m/min	50	100	200	–
Feed rate (f)	mm/rev	0.05	0.10	0.15	0.20

Adapted from Gaitonde et al. [6], with permission from SAGE

of each of the sources of variation by the respective *DF*. The Fisher's variance ratio is the ratio of *MS* of regression to *MS* of residual error. As per ANOVA, the model developed is adequate within the desired confidence interval, if Fisher's variance ratio of regression exceeds the standard tabulated value of *F*-ratio.

7.2.1.1 Case Study: RSM Model Development for Machining of MMC

The study presented here describes the development of second-order RSM-based models to predict some aspects of machinability, namely, machining force (F_m), cutting power (P) and specific cutting force (K_s) during turning of metal matrix composites (MMC) [6]. Aluminum alloy reinforced with 20% of silicon carbide (SiC) particulates (A356/20/SiCp-T6) work material was used throughout the investigation. The chemical composition of A356 aluminum matrix is with 7% Si and 0.4% Mg. The average dimension of SiC particles is about 20 μm.

Three levels for cutting speed (v) and four levels for feed rate (f) were selected and are given in Table 7.1. The turning experiments were planned as per full-factorial design (FFD) and were performed using MMC's appropriate workpieces with a diameter of 60 mm and a length of 200 mm using a PCD tool (*TCMW 16T3 04 FP CD10*). A '*STGCL 2020 K16*' type tool holder was used. The tool geometry was as follows: rake angle 0°, clearance angle 7°, cutting edge angle 91° and cutting edge inclination angle 0°. A '*Kingsbury MHP 50*' CNC lathe with 18 kW spindle power and a maximum spindle speed of 4,500 rpm was used to conduct the experiments. All the experiments were performed using a cutting fluid (emulsion 1/10 with BP Microtrend 231 L). Two millimeter depth of cut was kept constant throughout the investigation.

A Kistler® 9121 piezoelectric dynamometer with a charge amplifier (model 5019) was used to acquire three different components of forces, namely, cutting force (F_c), feed force (F_f) and depth force (F_d). The data acquisition was made through charge amplifier and a computer using appropriate software (Dynoware by Kistler®). The machining force, cutting power and specific cutting force are computed as:

$$F_m = \sqrt{F_c^2 + F_f^2 + F_d^2} \tag{7.5}$$

$$P = F_c v \tag{7.6}$$

Table 7.2 Experimental layout plan and the machinability characteristics

Trial number	Levels of process parameters		Actual values of process parameters		Machinability characteristics		
	v	f	v (m/min)	f (mm/rev)	F_m(N)	P(W)	K_s (MPa)
1	1	1	50	0.05	216.2	145	1,744.3
2	1	2	50	0.1	326.9	224	1,346.2
3	1	3	50	0.15	440.2	305	1,221.0
4	1	4	50	0.2	551.3	385	1,156.0
5	2	1	100	0.05	220.3	274	1,643.4
6	2	2	100	0.1	328.8	433	1,299.1
7	2	3	100	0.15	429.5	582	1,163.2
8	2	4	100	0.2	515.9	717	1,074.8
9	3	1	200	0.05	212.8	522	1,565.5
10	3	2	200	0.1	309.0	812	1,218.1
11	3	3	200	0.15	389.8	1,071	1,071.2
12	3	4	200	0.2	461.5	1,308	980.9

Adapted from Gaitonde et al. [6], with permission from SAGE

$$K_s = \frac{F_c}{f * d} \tag{7.7}$$

where, d is the depth of cut. The corresponding experimental layout plan along with the responses is presented in Table 7.2.

Following the model development procedure as explained in Sect. 7.2.1, the second-order RSM-based models of the machinability characteristics in natural form as reported by Gaitonde et al. [6] are given by:

$$F_m = 71.26667 + 0.376571\,v + 2782.8f - 3.79257143\,vf - 0.0006867\,v^2 - 1540f^2 \tag{7.8}$$

$$P = -77.125 + 2.201429\,v + 1,096.333f + 24.00857143\,vf - 0.0031167\,v^2 - 2,533.33333f^2 \tag{7.9}$$

$$K_s = 2,275.417 - 1.97179\,v - 10,772f - 0.28971429\,vf + 0.00382\,v^2 + 28,203.3333f^2 \tag{7.10}$$

where, v is in m/min; f in mm/rev; F_m in N, P in W and K_s in MPa.

The model adequacy is checked through ANOVA and is summarized in Table 7.3. Here, the developed mathematical models of machinability characteristics are significant at 95% confidence interval as F-ratio of all the models is >4.39 (F-table $_{(5,\ 6,\ 0.05)}$).

The proposed RSM-based mathematical models of machinability characteristics are used to analyze the interaction effects of process parameters by substituting the values of cutting speed and feed rate within the ranges selected [6]. The interaction

Table 7.3 Summary of ANOVA for machining force, cutting power and specific cutting force models

Response	Sum of squares		Degrees of freedom		Mean square		F-ratio
	Regression	Residual	Regression	Residual	Regression	Residual	
Machining force (Fm)	1,49,186	136	5	6	29,837	23	1,312.00
Cutting power (P)	13,88,586	536	5	6	2,77,717	89	3,111.35
Specific cutting force (K_s)	6,38,456	5,657	5	6	1,27,691	243	135.44

Adapted from Gaitonde et al. [6], with permission from SAGE
$F\text{-table}_{(5,\ 6,\ 0.05)} = 4.39$

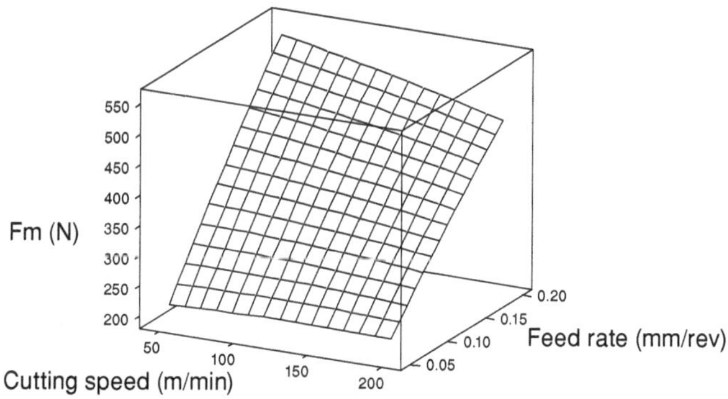

Fig. 7.1 Effect of cutting speed and feed rate on machining force (adapted from Gaitonde et al. [6], with permission from SAGE)

effects of cutting speed and feed rate on machining force, cutting power and specific cutting force as reported by Gaitonde et al. [6] are illustrated in Figs. 7.1, 7.2 and 7.3. It is observed from Fig. 7.1 that for any value of cutting speed, machining force increases with feed rate during turning of MMCs. Further, the machining force decreases with increase in cutting speed and the machining force is highly sensitive to variations in cutting speed at higher values of feed rate as compared to lower values. As seen from Fig. 7.2, the cutting power increases with feed rate for any given value of cutting speed during turning of MMCs and the cutting power is sensitive to variations in cutting speed at higher values of feed rate as compared to lower values. As depicted in Fig. 7.3, for a given value of cutting speed, the specific cutting force decreases with increase in feed rate and for a given feed rate, the specific cutting force is less sensitive to cutting speed variations during MMCs machining. A combination of higher feed rate with high cutting speed is found to be beneficial for minimizing specific cutting force.

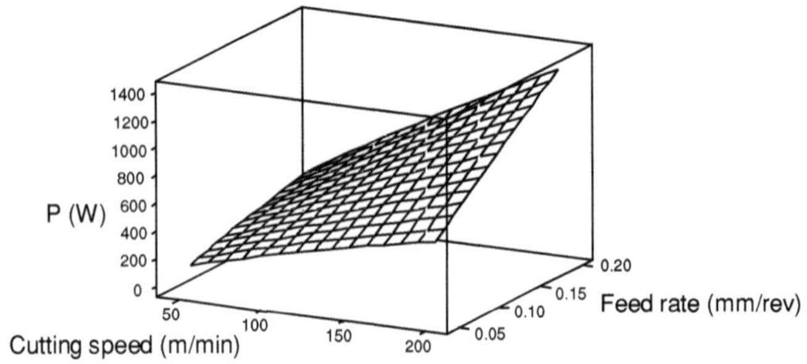

Fig. 7.2 Effect of cutting speed and feed rate on cutting power (adapted from Gaitonde et al. [6], with permission from SAGE)

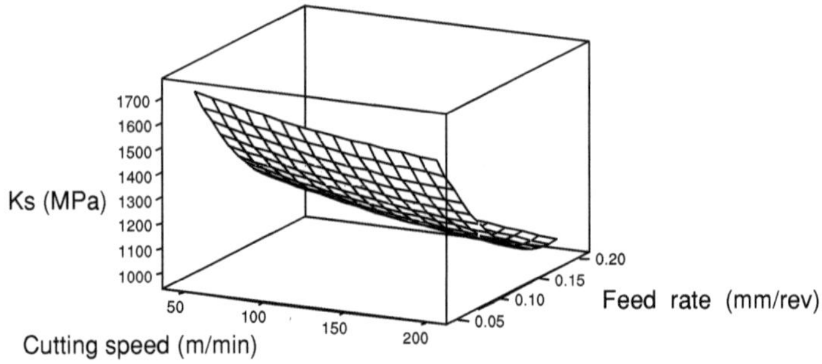

Fig. 7.3 Effect of cutting speed and feed rate on specific cutting force (adapted from Gaitonde et al. [6], with permission from SAGE)

7.2.2 Modeling Based on Artificial Neural Networks

The model development by RSM is a method, which requires minimum number of experiments to be conducted, but restricted to only small range of input variables and hence not suitable for complex and highly nonlinear processes. On the other hand, the development of higher order RSM model requires more number of experiments to be performed and hence costlier. This poses a limitation on the use of RSM models for highly nonlinear process and these constraints led to the development of model based on ANN.

The ANN is a fast, efficient, accurate and cost effective process-modeling tool in which the biological neurons are represented by a mathematical model [7]. The ability of ANN to capture any complex input–output relationships from the limited data set is valuable in machining process, where a huge experimental data for process-modeling is difficult and further expensive to obtain. The ANN is a flexible modeling tool with an aptitude to learn the mapping between input and output variables [7].

The purpose of ANN development is to imitate human brain so as to implement the various functions such as association, self-organization and generalization. The ANNs are parallel computer models of processes and the mechanisms, which constitute biological nerve systems. The ANNs are attractive in view of their high-execution speed and modest computer hardware requirements in addition to an adaptive nature and capability in solving complex and nonlinear problems. The ANN is made up of neurons connected via links. The information is processed within the neurons and is propagated to other neurons through the links connecting neurons.

Normally, a multi-layer feed forward ANN using error back propagation training algorithm (EBPTA) has been adapted. The EBPTA is a supervised learning algorithm based on generalized delta rule [7], which requires a set of inputs and desired outputs, known as training patterns. The EBPTA uses a gradient search technique, which updates the synaptic weights of connecting links during learning stage in such a way that mean square error (*MSE*) between the actual and desired output responses is minimized. The multi-layer feed forward ANN consists of neurons divided into input layer, hidden layer(s) and the output layer. The net activation input for *i*th neuron is given by [7]:

$$net_i = \sum_{j=1}^{n} w_{ij} x_i \tag{7.11}$$

where, w_{ij} = weight of link connecting neuron i to j; x_i = the output of ith neuron. For an unipolar sigmoid transfer function, the ouput of ith neuron is given as [7]:

$$o_{\mathrm{i}} = \frac{1}{1 + e^{\eta \, net_{\mathrm{i}}}} \tag{7.12}$$

where, η is the scaling factor. The training algorithm used here is based on the weight updates so as to minimize the sum of squared error for K number of neurons in the output layer and is given by [7]:

$$E = \frac{1}{2} \sum_{\mathrm{k}=1}^{K} \left(d_{\mathrm{k,p}} - o_{\mathrm{k,p}} \right)^2 \tag{7.13}$$

where, $d_{\mathrm{k,p}}$ = desired output for pth pattern. The synaptic weights of the links are updated as [7]:

$$w_{ji(\mathrm{n}+1)} = w_{ji(\mathrm{n})} + \alpha \, \delta_{pj} o_{pi} + \beta \, \Delta w_{ji(\mathrm{n})} \tag{7.14}$$

where, n is the learning step, α is the learning rate and β is the momentum constant. The error term δ_{pj} is given by:

- For output layer:

$$\delta_{pk} = (d_{kp} - o_{kp})(1 - o_{kp}); \; k = 1, \ldots K \tag{7.15}$$

- For hidden layer:

$$\delta_{pj} = o_{pj}(1 - o_{pj}) \sum \delta_{pk} w_{kj};\ j = 1, \ldots J \tag{7.16}$$

where, J is the number of neurons in the hidden layer.

The steps involved in ANN training using EBPTA are mentioned below:

1. The network weights are initialized to small random values.
2. The input/desired output pairs are presented one by one, updating the weights each time.
3. The *MSE* due to all outputs and *NP* number of patterns is computed as:

$$MSE = \frac{1}{NP} \sum_{p=1}^{NP} \sum_{k=1}^{K} \left(d_{kp} - o_{kp}\right)^2 \tag{7.17}$$

4. If ($MSE <$ specified tolerance) or (epochs $>$ $(\text{epochs})_{\max}$)

   Then stop.
   Else, go to Step 2.

7.2.2.1 Case Study: ANN Model Development for Machining of MMC

A case study of ANN-based modeling of surface roughness in turning of Al-SiC (20p) using coarse grade polycrystalline diamond (PCD) inserted under different cutting conditions is considered in this section [8]. A multi-layer perceptron (MLP) model has been constructed with EBPTA to capture the relationship between cutting speed (v), feed rate (f) and depth of cut (d) on surface roughness (R_a) of turned component. The input–output data required for development of ANN model has been obtained through FFD. The experimental results were obtained by turning of MMC's of type A356/SiC/20p (aluminum with 7.5% silicon, 2.44% magnesium, reinforced with 20% volume particles of SiC).

A medium duty lathe of 2 kW spindle power has been employed for dry turning of 30 trials of parameter combinations. The CNMA 120408 inserts with PCLNR 25 X25 M12 tool holder with PCD were used to turn the billets of 150 mm diameter. The tool geometry of PCD inserts employed was, top rake angle of 0° and nose radius of 0.8 mm. The work material was machined at five different cutting speeds ranging from 100 to 600 m/min with two feed rates of 0.108 and 0.200 mm/rev and depth of cut as 0.25, 0.50 and 0.75 mm. Each experimental trial was carried out for 3 min duration.

The average surface roughness (R_a) in the direction of tool movement was measured in three different places of machined surface using a surface roughness tester, Mitutoyo Surf test-301 with a cut-off and transverse length of 0.8, and 2.5 mm, respectively. The average surface roughness (R_a) for three different locations was considered for each trial.

Table 7.4 Validation of results for surface roughness obtained using ANN

Reading number	Experimental surface roughness (μm)	ANN predicted surface roughness (μm)	Error (%)
1	3.94	3.96	0.75
2	2.27	2.25	0.50
3	4.21	4.30	2.17
4	3.87	3.88	0.27
5	4.50	4.45	1.10
6	2.49	2.52	1.42
7	6.19	6.10	1.44
8	5.05	5.03	0.39
9	5.39	5.13	4.78
10	2.93	2.86	2.26
11	5.75	5.66	1.48
12	4.57	4.60	0.84

Adapted from Muthukrishnan and Davim [8], with permission from Elsevier

The ANN training was performed using 18 input–output patterns and other 12 data sets were then utilized for ANN validation. The network was trained by using suitable scaling factor for input parameters. The ANN designed for the present study takes depth of cut, cutting speed and feed rate as the input parameters; surface roughness as the output parameter. The ANN architecture selected for the surface roughness force model is 3-10-1, number of epochs is set to 2000, transfer function is sigmoid, learning factor is 0.6 while momentum factor is 1.0.

The average error when testing all training and testing patterns was found to be 1.47% for the developed ANN-based surface roughness model. The validation of results for surface roughness obtained using ANN is presented in Table 7.4. The details of ANN training and model verification are given in Muthukrishnan and Davim [8]. The proposed ANN model of surface roughness can be used to analyze the interaction effects of process parameters on surface roughness of turned components of MMC by generating 3D surface plots.

7.3 Optimization Methods

The robust parameter design (RPD) as suggested by Taguchi [2, 4] is an engineering methodology that emphasizes proper choice of levels of control factors in a process. The principle of choice of levels focuses mainly on variability around a target for the quality characteristic. The majority of variability around a target is caused by uncontrollable parameters known as noise factors. Hence, RPD entails designing the system by selecting the optimal levels of control factors so as to achieve robustness of system response to inevitable changes in the noise factors [2, 4]. The noise factors may be, and often are, controlled at research or development level but they cannot be controlled at the production or product use level [4].

Traditionally, mathematical programing techniques like linear programing, integer programing, dynamic programing and geometric programing have been used to solve the optimization problems in machining processes. However, the traditional techniques are not ideal for solving these problems, as they tend to obtain a local optimal solution. Considering the drawbacks of the traditional optimization techniques, attempts are being made to optimize the machining problem using heuristic search algorithms like GA, SA, ACO and PSO [5].

7.3.1 Taguchi Robust Design

Taguchi-based optimization technique has produced a unique and powerful optimization discipline that differs from the traditional practises. Taguchi optimization technique is based on concept of "robust design", which aims at obtaining the solutions that make the design less sensitive to noise factors. Taguchi technique [4] has been widely applied in the process design, wherein the mathematical models for the performance do not exist and the experiments are typically conducted to determine the optimum settings for the design and process variables.

Taguchi robust design principle is based on matrix experiments. The traditional experimental design methods are too complex and are not easy to use. If the number of parameters is more, a large number of experiments have to be performed. This problem is overcome in the Taguchi technique, which uses a special design of orthogonal arrays (OA) [4, 9] to study the entire parameter space with small number of experiments. The OA is a major tool used in the robust design, which is used to study many design parameters by means of a quality characteristic. The purpose of conducting an experiment based on OA is to determine the optimum level for each parameter and to establish the relative significance of individual parameters in terms of their main effects on the quality characteristic. The OA gives acceptable estimates of factor effects with reduced number of experiments when compared to the traditional methods.

Depending on number of factors and the levels defined for each of the factors, a suitable OA is selected. Each column of OA designates a factor and its setting levels in each experiment and each row designates an experiment trial with combination of levels of different factors in that trial. The first step in constructing an OA to fit a specific case study is to count the total DF, which gives the minimum number of experiments to be conducted. The selection of OA [9] begins with the distinct number of levels (l) defined for the number of factors (k). The minimum number of trials in the OA is:

$$N_{\min} = (l - 1)k + 1 \tag{7.18}$$

Taguchi has tabulated 18 standard OAs, each comprising N_0 ($\geq N_{\min}$) number of trials for different numbers of factors and their levels, which can be directly used for the experimental plan [9]. However, Taguchi design allows defining different number of levels for each factor. In such situations, the mixed level OA need to be selected for the experimentation purpose. After a simple analysis and processing of the output results from experiments as per OA, an optimum combination of the factor values can be obtained. It is demonstrated in statistics that although the number of experiments is dramatically reduced, the optimal result obtained from OA usage is very close to that obtained from FFD.

The principle behind Taguchi robust design is to control the effect of variations caused by noise factors on product quality characteristic rather than controlling the source of noise itself [4, 9]. In order to minimize the variations in the quality characteristic, Taguchi introduced a method to transform the repetition data to signal-to-noise (S/N) ratio (η), which is a measure of variation present in the scattered response data [4, 9]. The maximization of S/N ratio simultaneously optimizes the quality characteristic and minimizes the effect of noise factors.

For each trial "i" in the OA, if the performance measure (y) is repeated "n" times, then S/N ratio can be computed as follows [4, 9]:

- Smaller-the-better type:

$$\eta_i = -10\log_{10}\left[\frac{1}{n}\sum_{j=1}^{n} y_j^2\right] \text{ dB} \tag{7.19}$$

if y needs to be minimized.

- Larger-the-better type:

$$\eta_i = -10\log_{10}\left[\frac{1}{n}\sum_{j=1}^{n} y_j^{-2}\right] \text{ dB} \tag{7.20}$$

if y needs to be maximized.

Taguchi optimization procedure consists of analysis of means (ANOM) and ANOVA on S/N ratio of OA [4]. The ANOM is used to identify the optimal factor level combinations and to estimate the main effects of each factor. ANOM is also employed to find the effect of a factor level, which is the deviation it causes from the overall mean response. The optimal level for a parameter is the level, which results in highest value of S/N ratio in the experimental region. The ANOVA is used to estimate the error variance for the effects and variance of the prediction error. ANOVA is performed on S/N ratio to obtain the contribution of each of the factors.

After selecting the optimal levels of process parameters, the final step is to predict and verify the adequacy of the model for determining the optimum response [4, 9]. The confirmation experiments under the optimal conditions are

Table 7.5 Results of ANOVA of surface roughness

Cutting parameters	Degrees of freedom	Sum of squares	Mean square	F-test	Percent contribution
Cutting speed	2	60.70	30.38	28	12
Feed rate	2	35.70	17.90	6.2	51
Depth of cut	2	16.90	8.40	18.80	30
Error	20	45.20	1.125	–	7
Total	26	158.50	–	–	100

Adapted from Muthukrishnan and Davim [8], with permission from Elsevier

then performed and the results with the predictions are compared. In order to judge the closeness of observed value of signal-to-noise ratio with that of the predicted value, the variance of prediction error is determined and the corresponding two-standard deviation confidence limits for the prediction error of S/N ratio are calculated. If the prediction error is outside these limits, one should suspect the possibility that the additive model is not adequate. Otherwise, the additive model is adequate [4, 9].

7.3.1.1 Case Study: Taguchi Robust Design for Machining of MMC

This case study demonstrates the application of Taguchi method to determine the optimal process parameter settings, namely, cutting speed (v), feed rate (f) and depth of cut (d) in order to minimize the surface roughness (R_a) during turning of Al-SiC (20p) MMC using coarse grade PCD insert [8]. As in previous case study explained in Sect. 7.2.2.1., the same workpiece material, cutting tool and the experimental set up were employed in the current investigation. The experiments are planned as per Taguchi's L_{27} orthogonal array and each trial of an array consists of three replications. The S/N ratio for the selected performance characteristic (smaller-the-better type) is given by [8]:

$$\eta = -10\log_{10}\left[\frac{1}{r}\sum_{i=1}^{r} R_i^2\right] \tag{7.21}$$

where, R_i is the value of the surface roughness for the test in that trial and r is the number of tests in a trial. For lower-the-better characteristics, this translates into lower process average and improved consistency from one unit to the next or both.

The ANOM gives the optimal levels of the process parameter combination and the ANOVA summarizes the percent contribution of each factor. The optimum factor level combinations obtained through ANOM reported by Muthukrishnan and Davim [8] are cutting speed at 575 m/min, feed rate at 0.108 mm/rev and depth of cut at 0.75 mm. The results of ANOVA for surface roughness are given in Table 7.5. The details of main effects of process parameters, the prediction and verification of quality characteristic using the optimal level of the design parameters are presented in Muthukrishnan and Davim [8]. It was reported that increase in S/N ratio from initial cutting parameters is 3.99 dB for surface roughness, which implies that the surface roughness qualities have improved. It was also observed in their study that the experimental results were close to the predicted values and were falling within the confidence limits.

7.3.2 Heuristic Search Algorithms

The heuristic algorithms are the model-based optimization techniques i.e., by using the metamodels to obtain predictions of the phenomena of interest and then to find the best compromises among the objectives for the system through the latest non-traditional optimization tools like GA, SA, ACO and PSO.

7.3.2.1 Genetic Algorithms

The GA are non-traditional search algorithms that emulate the adaptive processes of natural biological systems [10]. Based on the survival and reproduction of the fittest, they continually search for new and better solutions without any pre-assumptions such as continuity and unimodality. The GA has been applied in many complex optimization and search problems, outperforming the traditional optimization and search methods.

The solution of the problem that GAs attempt to solve is coded into a string of binary numbers known as chromosomes. Each chromosome contains the information of a set of possible process parameters. Initially, a population of chromosomes are formed randomly. The fitness of each chromosome is then evaluated using an objective function after the chromosome has been decoded. Upon completion of the evaluation, either a roulette wheel method or selected control method is used to select randomly pairs of chromosomes to undergo genetic operations such as crossover and mutation to produce offspring for fitness evaluation. This process continues until a near optimal solution is found.

7.3.2.2 Simulated Annealing

The SA is also one of the non-traditional search and optimization techniques, which resembles the cooling process of molten metals through annealing [11]. The SA procedure simulates this process of annealing to achieve the minimization function value in a problem.

The algorithm begins with an initial point, m_1 and a high temperature, T. A second point, m_2 is created using a Gaussian distribution and the difference in the function values at these points (ΔE), is calculated. If the second point has a smaller value, the point is accepted; otherwise the point is accepted with a probability $e^{(-E/T)}$ [11]. This completes an iteration of the SA procedure.

The algorithm is terminated when a sufficiently small temperature is obtained or a small enough change in the function value is observed.

7.3.2.3 Ant Colony Algorithm

The natural metaphor on which ant algorithms are based is ant colonies. The researchers are fascinated by seeing the ability of near-blind ants in establishing the shortest route from their nest to the food source and back. These ants secrete a substance, called pheromone, and use its trial as a medium for communicating information [12]. The probability of the trial being followed by other ants is enhanced by further deposition by others following the trial. This cooperative behavior of ants inspired the new computational paradigm for optimizing real-life systems, which is suited for solving large-scale problems.

In the first step of ant colony algorithm (ACO), hundred solutions are generated randomly with parameters that satisfy the constraints. The initial solutions are classified as superior and inferior solutions. The following three operations are performed on the randomly generated initial solution: (1) random walk or cross over—90% of the solutions (randomly chosen) in the inferior solutions are replaced with randomly selected superior solutions, (2) mutation—the process whereby randomly adding or subtracting a value is done to each variable of the newly created solutions in the inferior region with a mutation probability and (3) trial diffusion—applied to inferior solutions that were not considered during random walk and mutation stages.

7.3.2.4 Particle Swarm Optimization

The PSO is a population-based stochastic optimization technique, inspired by social behavior of bird flocking or fish schooling [13]. The PSO shares many similarities with evolutionary computation techniques such as GA. The system is initialized with a population of random solutions and searches for optima by updating generations. However, unlike GA, PSO has no evolution operators such as crossover and mutation.

In PSO, the potential solutions, called particles, fly through the problem space by following current optimum particles. Each particle keeps track of its coordinates in the problem space, which are associated with the best solution (fitness) it has achieved so far and the fitness value is stored. This value is called "pbest". Another "best" value that is tracked by the particle swarm optimizer is the best value, obtained so far by any particle in the neighbors of the particle. This location is called "lbest". When a particle takes all the population as its topological neighbors, the best value is a global best and is called "gbest". The PSO concept consists of, at each time step, changing the velocity of (accelerating) each particle toward its "pbest" and "lbest" locations (local version of PSO). Acceleration is

Table 7.6 Turning cutting conditions with PCD

S. No.	Cutting speed (m/min)	Feed rate (mm/rev)
1	250	0.1
2	350	0.1
3	500	0.1
4	700	0.1
5	500	0.2
6	500	0.05

Adapted from Antonio and Davim [14], with permission from Elsevier

weighted by a random term, with separate random numbers being generated for acceleration toward "pbest" and "lbest" locations.

7.3.2.5 Case Study: Genetic Algorithms-Based Optimization for Machining of MMC

This case study addresses the application of GA for optimizing the cutting conditions during turning of PMMC's of type A356/20/SiCp-T6 (continuous casting) in which the matrix is aluminum with 7% silicon, 0.4% magnesium, reinforced with 20% volume particles of SiC [14]. The material was subjected to heat treatment (solutionizing and aging T6-5h at 154°C). The average dimension of SiC particle is 20 microns.

A lathe with 6 kW spindle power lathe was employed for the current investigation. TMCW 16T308F PCD inserts were used for machining billets of 95 mm diameter lubricated with an emulsion (Alusol—B 8%). Constant depth of cut of 1 mm was employed in the study. The turning conditions with PCD inserts are presented in Table 7.6.

A Kistler piezoelectric dynamometer with appropriate load amplifier was used. Different programs for data acquisition have been developed and used based on Lab VIEW software. They allow continuous recording and simultaneous graphical visualization of evolution of cutting force, feed force and depth force. The tool wear was measured according to ISO 3685 with a Mitutoyo optical microscope, which has 30× magnification and 1 micron resolution. The surface roughness of turned workpiece was evaluated according to ISO 4287/1 with a HomeltesterT500 profilometer. The details of experimental results and discussion are given in Antonio and Davim [14].

The aim of the numerical model was to obtain the cutting conditions using GA. The controlled variables cutting speed (v), feed (f) and cutting time (t) assume the following discrete values: $v = (v_1, v_2, \ldots, v_n); f = (f_1, f_2, \ldots, f_n)$ and $t = (t_1, t_2, \ldots, t_n)$ with a genetic code to define. As reported in Antonio and Davim [14], the time interval is 39 min with fractions of 1 min and hence the design space is a typical discrete and non-convex search domain. Each chromosome has three genes and each gene represents code value for each variable on a turning operation according to above discrete values. The various outputs considered in this study are cutting force (F_c), feed force (F_f), depth force (F_d), tool wear (V_b), average surface roughness (R_a) and peak to valley height (R_t).

Denoting the turning parameters by U_i, the normalization with respect to maximum values is performed as:

$$U_i^* = \frac{U_i}{U_i^{\max}}, \; U_i^{\max} = Maximum\left(U_i^j; j = 1, \ldots, \bar{N}\right) \tag{7.22}$$

where, *j*th superscript refers to individual experiments and $\bar{N}$ is the total number of experimental values. Here, the given problem is multi-objective optimization, which involves minimization of several machining performance measures. The total utility function in the current study is given by:

$$\bar{U}(v, f, t) = \sum_{i=1}^{6} \omega_i U_i^* \tag{7.23}$$

where, ω_i is scalar weighting factor associated with *i*th objective. As stated in Antonio and Davim [14], the present problem is a multi-criteria optimization with contradictory objectives and the formulation of the problem is given by:

$$\text{Maximize, } F = \frac{t^*}{\bar{U}(v, f, t)} \tag{7.24}$$

Subject to constraints, $V_b < 0.3$ mm; $R_a < 1$ μm.

Where, $t^* = \frac{t}{t^{\max}}$is the normalized cutting time.

The fitness function considers the above constraints, which can be applied via penalty functions. The application of penalties to individuals where penalties are violated is carried out by:

$$U_i^{pen} = U_i + kd^n \tag{7.25}$$

where, U_i is the unpenalized fitness, $U_i^{pen}$is the penalized fitness, *d* is the difference between actual and allowable values of design constraint and *k*, *n* are constants to be determined [14]. The effectiveness of penalties relates to the comparison with magnitude of maximum fitness $U_i^{\max}$in Eq. 7.22. To avoid some heavily penalised individuals remaining in the population, special attention is given to values of constraint violation (d_0), which can be tolerated, and lowest value of difference (d_1) attracting a severe penalty. If p_0 and p_1 are the penalties corresponding to d_0 and d_1 then,

$$p_0 = \varepsilon_0 U_i^{\max} \tag{7.26}$$

$$p_1 = \varepsilon_1 U_i^{\max} \tag{7.27}$$

where, ε_i is the percentage of maximum value of cutting parameter.

The fitness function corresponding to current optimization problem is given by:

$$FIT = \frac{t^*}{\omega_1 F_c^* + \omega_2 F_f^* + \omega_3 F_d^* + \omega_4 (V_b^*)^{pen} + \omega_5 (R_a^*)^{pen} + \omega_6 (R_t^*)} \tag{7.28}$$

where $(V_b^*)^{pen}$ and $(R_a^*)^{pen}$ are the penalized cutting parameters related to degree of violation of imposed constraints obtained from Eq. 7.25 and the normalization proposed in Eq. 7.22. Here, the equal contributions from each cutting parameter are considered in the fitness function Eq. 7.28.

The details of genetic operations, namely, selection, crossover and mutation are given in Antonio and Davim [14]. A population of eight chromosomes are considered in GA and search is based on best solutions with $N_A = 2$ and $N_c = 1$ is considered for mutation operator. The constraints are penalized by $\varepsilon_0 = 1\%$ and $\varepsilon_1 = 5\%$. The constraint violations considered are: For V_b, $d_0 = 0.003$ mm, $d_1 = 0.03$ mm; For R_a, $d_0 = 0.05$ microns, $d_1 = 0.1$ microns. The optimal cutting conditions are found to be cutting speed (v) = 350 m/mi, feed (f) = 0.1 mm/rev and cutting time (t) = 19 min. It was concluded that the optimal cutting conditions of turning operation obtained through GA optimization demonstrate the potential of mixed numerical-experimental mode.

7.4 Conclusion

The modeling based on RSM and ANN have been extensively used in the machining processes. The RSM coupled with DOE is a collection of mathematical and statistical technique not only reduces the cost and time, but also gives the required information about the main and interaction effects of the factors with minimum number of experiments. On the other hand, ANN is a powerful modeling tool; mainly deals with complex and nonlinear problems and can provide accurate results in machining process.

Two case studies based on modeling approaches involving the machinability investigations during turning of MMC with PCD tool were presented. The first investigation is on RSM modeling to study the influence of cutting speed and feed on machining force, cutting power and specific cutting force. The two-factor interaction effects on machinability characteristics were studied by generating 3D response surface plots. The ANN-based modeling for predicting the surface roughness during turning of MMC is detailed in the second case study. A MLP model trained by EBPTA has been used to capture the relationship between cutting speed, feed and depth of cut on surface roughness.

The main advantage of Taguchi robust design (TRD) is to make the process performance measure less sensitive to noise factors. The TRD employs OA for conducting the experiments and signal-to-noise (S/N) ratio as the objective function for optimization. In Taguchi method, analysis of means (ANOM) is used to identify the optimum level factor combinations and ANOVA is employed to estimate the relative significance of each factor on performance measure. The case study demonstrating the application of Taguchi method for determining the optimal process parameter settings of cutting speed, feed rate and depth of cut to minimize the surface roughness during turning of MMC is presented in this chapter.

The conventional optimization techniques such as gradient-based methods do not function effectively and are not suitable for solving multi-objective optimization problems. Hence, a number of heuristic algorithms such as GA, SA, ACO and PSO have been proposed for obtaining optimal solutions for multi-objective problems. The multi-objective optimization in machining of MMC using GA concept is illustrated in this chapter, which consists of a stochastic strategy of direct search that stimulates the genetic process of evolution.

Acknowledgements The authors would like to thank Elsevier and SAGE publications for granting permission for re-use of the published materials.

References

1. Montgomery DC (2004) Design and analysis of experiments. Wiley, New York
2. Myers RH, Montgomery DC, Anderson-Cook CM (2009) Response surface methodology. Wiley, New Jersey
3. Gaitonde VN, Karnik SR, Davim JP (2009) Design of experiments. In: Ozel T, Davim J (eds) Intelligent machining: modeling and optimization of the machining processes and systems. Wiley, USA, pp 215–243
4. Phadke MS (1989) Quality engineering using robust design. Prentice Hall, Englewood Cliffs, NJ
5. Satishkumar S, Asokan P, Kumanan S (2006) Optimization of depth of cut in multi-pass turning using nontraditional optimization techniques. Int J Adv Manuf Technol 29:230–238
6. Gaitonde VN, Karnik SR, Davim JP (2009) Some studies in metal matrix composites machining using response surface methodology. J Reinforc Plast Compos 28(20):2445–2457
7. Schalkoff GB (1997) Artificial neural network. McGraw-Hill, Singapore
8. Muthukrishnan N, Davim JP (2009) Optimization of machining parameters of Al/SiC-MMC with ANOVA and ANN analysis. J Mater Process Technol 209:225–232
9. Ross PJ (1996) Taguchi techniques for quality engineering. McGraw-Hill, Singapore
10. Goldberg DE (1989) Genetic algorithms in search optimization and machine learning. Addison-Wesley, New York
11. Deb K (1995) Optimization for engineering design: algorithms and examples. Prentice-Hall, New York
12. Dorigo M (1996) The ant system: optimization by a colony of cooperating agent. IEEE Trans Syst Man Cybern Part B 26(1):1–13
13. Kennedy J, Eberhart RC (1995) Particle swarm optimization. In: Proceedings of IEEE international conference on neural networks. University of Western Australia, Perth, Western Australia, pp 1942–1948
14. Antonio CAC, Davim JP (2002) Optimal cutting conditions in turning of particulate metal matrix composites based on experiment and a genetic search model. Composites Part A 33:213–219

Index

J. P. Davim (ed.), *Machining of Metal Matrix Composites*,
DOI: 10.1007/978-0-85729-938-3,

MIX
Papier aus verantwortungsvollen Quellen
Paper from responsible sources
FSC® C105338

If you have any concerns about our products,
you can contact us on
ProductSafety@springernature.com

In case Publisher is established outside the EU,
the EU authorized representative is:
Springer Nature Customer Service Center GmbH
Europaplatz 3, 69115 Heidelberg, Germany

Printed by Libri Plureos GmbH
in Hamburg, Germany